AQUEOUS-ENVIRONMENTAL CHEMISTRY OF METALS

AQUEOUS-ENVIRONMENTAL CHEMISTRY OF METALS

ALAN J. RUBIN, Editor

Professor of Civil Engineering
Water Resources Center
The Ohio State University
Columbus, Ohio

ann arbor science PUBLISHERS INC.
POST OFFICE BOX 1425 • ANN ARBOR, MICHIGAN 48106

Second Printing, 1976

Copyright © 1974 by Ann Arbor Science Publishers, Inc.
P.O. Box 1425, Ann Arbor, Michigan 48106

Library of Congress Catalog Card Number 74-78805
ISBN 0-250-40060-X

Manufactured in the United States of America
All Rights Reserved

PREFACE

This book deals with the chemistry of metals in aqueous systems. The specific metals discussed are those of environmental interest either because of their health implications or because of their significance in water and wastewater treatment. The tone for the book is set in the first paper with subsequent chapters detailing or expanding on the points raised.

Chapter 1 by Leckie and James reviews the various interactions of metal species. Consideration is given to chemical systems, such as those of sulfide and carbonate, that control metal concentration, form and distribution. Reactions and mechanisms for metal transport, especially through the sediments, are explained using mercury(II) as the example. Some discussion of the sources of metals to natural aquatic systems is also given. This is discussed in greater detail by Williams, Aulenbach and Clesceri in Chapter 2. Data from the literature and their own investigations are presented to demonstrate the differences in the distribution of metals in streams and in oligotrophic and eutrophic lakes. Schell and Barnes in the book's third chapter describe a case study of the distribution of lead and mercury in the waters, sediments and biota of western Washington State.

Chapter 4, by Brezonik, reviews some aspects of the water quality significance of metals and methods for their analysis at the trace level. Applications of kinetic analysis using the AutoAnalyzer and anodic stripping voltammetry for the determination and speciation of cadmium are discussed at length. An investigation of the oxidation of iron(II) by oxygen in buffered waters is described by Ghosh in Chapter 5. The kinetics of the reaction and the applications of continuous-stirred-treatment-reactor theory are discussed. The results of an extensive study on the kinetics of crystal growth for calcium carbonate, magnesium hydroxide and dicalcium phosphate are presented in the next chapter by Nancollas and Reddy.

Cadmium is the subject of the following two chapters. Weber and Posselt describe solution equilibria including precipitation reactions in Chapter 7, whereas Chapter 8 is devoted to the oxidation of the solid metal and its subsequent entry to distribution systems. The book's final chapter, by Hayden and Rubin, describes an investigation of aluminum(III) hydrolysis and precipitation. Species and their formation constants, including solubility products for freshly prepared and aged aluminum hydroxide, were determined over a broad range of solution pH and metal ion concentrations. Solution conditions controlling the colloidal stability of the hydroxide are also examined.

The editor is grateful for the patience and cooperation of the contributors and the staff of the publisher. The support of the Civil Engineering Department and the College of Engineering of the Ohio State University is also gratefully acknowledged.

> Alan J. Rubin
> Columbus, Ohio
> August, 1974

CONTENTS

Chapter	Page
1. Control Mechanisms for Trace Metals in Natural Waters *James O. Leckie and Robert O. James*	1
2. Sources and Distribution of Trace Metals in Aquatic Environments *Sherman L. Williams, Donald B. Aulenbach, and Nicholas L. Clesceri*	77
3. Lead and Mercury in the Aquatic Environment of Western Washington State *W. R. Schell and R. S. Barnes*	129
4. Analysis and Speciation of Trace Metals in Water Supplies *Patrick L. Brezonik*	167
5. Oxygenation of Ferrous Iron(II) in Highly Buffered Waters *Mriganka M. Ghosh*	193
6. Crystal Growth Kinetics of Minerals Encountered in Water Treatment Processes *George H. Nancollas and Michael M. Reddy*	219
7. Equilibrium Models and Precipitation Reactions for Cadmium(II) *Walter J. Weber, Jr. and Hans S. Posselt*	255
8. Studies on the Aqueous Corrosion Chemistry of Cadmium *Hans S. Posselt and Walter J. Weber, Jr.*	291
9. Systematic Investigation of the Hydrolysis and Precipitation of Aluminum(III) *Phillip L. Hayden and Alan J. Rubin*	317
INDEX	383

CHAPTER 1

CONTROL MECHANISMS FOR TRACE METALS IN NATURAL WATERS

James O. Leckie
Civil Engineering Department
Stanford University, California

Robert O. James
Department of Physical Chemistry
University of Melbourne, Australia

Introduction	2
Sources of Trace Metals to Natural Waters	3
Natural Sources	3
Mining Operations and Fossil Fuels	5
Industrial and Domestic Wastes	6
Nature of Aqueous Metal Ions	7
Natural Waters as Electrolyte Solutions	7
Hydrolysis Reactions of Heavy Metals	11
Complexation of Metal Ions in Solution	16
Chelation with Organic Ligands	21
Oxidation–Reduction Reactions	23
Interactions in Heterogeneous Sulfide Systems	29
Interactions with Solid/Solution Interfaces	33
Nature of the Solid/Solution Interface	33
Adsorption Behavior of Trace Metals	36
Phenomenological Models for Interactions	46
General Model for Adsorption of Hydrolyzable Metal Ions	49
Transport Processes in Natural Aquatic Systems	63
Transport of Mercury	63
Relative Residence Time of Mercury	66
Temporal and Spatial Distribution of Mercury	68
Summary	69

INTRODUCTION

The environmental biogeochemist encounters a large number of phenomena which exert varying degrees of control on the behavior of trace metals in natural water systems. At present, the available chemical data and phenomenological models are not adequate to allow complete prediction of the distribution of trace metals in natural waters. These waters are highly complex electrolyte solutions in contact with a variety of inorganic and organic solids. Although trace metals enter natural water systems by normal weathering of minerals, significant localized inputs are made by man's activities. Whenever the rate of input of trace elements or their compounds into a water system exceeds the natural rate of cycling, contamination or adverse ecological effects may result.

Equilibrium solubility models are useful in establishing simple boundary conditions for the purpose of discussion. Consideration of the interaction between inorganic and organic species both in solution and at the solid/solution interface, however, appears to be necessary to account for many of the observed phenomena.

Apparent deviations from equilibrium model predictions are manifested in natural waters as supersaturation of some species with respect to their most stable solid compounds. In other cases, there is apparent undersaturation with respect to model compounds. In either case there is often little or no information concerning the nature of the species in solution nor is there much consideration of why the particular species exhibits anomalous behavior.

More information is needed on the types of species present in natural waters under a variety of different chemical conditions. The kind of data needed includes the type and number of complexes or compounds formed, their chemical stability and their rate of formation. Once these data are available, the distribution of metals among species in aqueous solutions can be handled with a great deal of sophistication by present methods. However, we need to improve our understanding of interfacial phenomena and apply our knowledge of the competitive equilibria in aqueous solutions to the interfacial regions bounding natural water systems. This kind of general treatment will be a prerequisite for insight into the distribution and control of trace elements in the environment.

In addition to interfacial processes being among the controls of the concentration of soluble metals, accumulation of heavy metals at the solid/solution interface may also be one of the important transport mechanisms for heavy metals in natural aquatic systems. The role of natural and synthetic organic solutes and colloids in the distribution of trace metals is complex and little understood. More research into the characterization of solution species involved in biogeochemical phenomena is required if eventually we are to elucidate and quantify the processes and mechanisms whereby trace metals are released, transported and accumulated or dispersed in natural aquatic systems.

Although the following discussion is not exhaustive, it attempts to give a general picture of the pertinent physical and chemical forms in which metals such as Hg, Pb, Zn and Cd appear, and the natural limits of their occurrence in natural water systems. Further, this paper reviews briefly those chemical properties and natural phenomena which, at present, are believed to influence the transport of these trace elements in natural waters.

SOURCES OF TRACE METALS TO NATURAL WATERS

Sources of trace metals are basically either natural (chemical weathering of rock, volcanic activity) or are man influenced (burning fossil fuels, mining operations, industrial uses of trace metals). Man's influence is essentially a kinetic one in that the rate or flux of trace metals in the general biogeochemical cycle is accelerated at specific points in the cycle due to man's activities (Figure 1.1). Specifically, activities in industrialized nations have accelerated the flux of trace metals in atmospheric and surficial environments; in the case of lead and mercury, the increased fluxes have been dramatic when viewed on a long term basis.

Natural Sources

On a geologic time scale chemical weathering and volcanic or tectonic activities have been the major release mechanisms

4 Aqueous-Environmental Chemistry of Metals

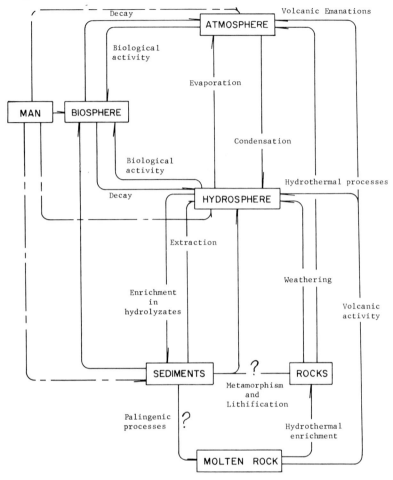

Figure 1.1. Biogeochemical cycle of mercury in nature with major input points of man's activities.

operating to introduce trace metals into surface waters. In the case of mercury, for example, volcanic activity can be responsible for major natural inputs in a localized sense. On the other hand, chemical weathering of igneous and metamorphic rock is responsible for the major natural flux of trace metals into surface aquatic environments. The role of chemical weathering reactions in producing the chemical composition of natural waters has been discussed extensively [1,2]. The significance of biochemical weathering is understood to a lesser degree. However, the role of natural organic chelating agents in the weathering process is significant even if unquantified [3].

In many urbanized regions and in areas where there has been considerable disruption of natural materials through excavations, strip mining and road building, the exposed and loose material undoubtedly weathers at an accelerated rate. The exposure of fresh rock to the atmosphere in highly mineralized mining regions is known to be one process which accelerates the mobilization of trace metals into aquatic systems.

Mining Operations and Fossil Fuels

Anomalous concentrations of trace metals in and near ore deposits are well known. The possibility of developing prospecting information from low level mineralization or halos of trace metals, especially mercury, has been known for some time. The exploitation of ore deposits invariably exposes large quantities of waste rock to accelerated weathering conditions. Rock in mine dumps, although usually too low grade to be of interest economically, is nevertheless still highly mineralized in trace metals in a relative sense when compared to background levels. In addition, ore processing and smelting techniques can sometimes mobilize large quantities of trace metals, for example, enhanced levels of total Cd, Pb, Ag, Zn and Cu in alluvial soils on old mining areas of Wales [4]. The pattern of contamination is apparently controlled by the composition of local ores.

The exact extent or significance of the enhanced flux due to mining and smelting operations is not yet known. It is known to be important on a local level where the flux is concentrated regionally. The case of cadmium in Japan is an example in the extreme where a number of casualties have been associated with the trace metal contamination [5].

Bertine and Goldberg [6] have estimated the flux of Pb and Hg from continental weathering to be 110×10^9 gm/yr and 2.5×10^9 gm/yr, respectively. Recent estimates indicate that, on a global scale, the present rate of input of trace metals such as Hg, Pb, Zn, Cd, as well as Ag, Cu, Sb, Sn, and others, is in excess of the natural rate of biogeochemical cycling [6-9]. Apparently, the input on an annual basis for Pb and Hg from fossil fuels alone is sufficient to make a significant difference in the global flux. The contribution due to burning fossil fuels is estimated to be 3.6×10^9 gm/yr and 1.6×10^9 gm/yr for Pb and Hg, respectively [6]. If the estimates of Bertine and Goldberg [6], and Joensuu [10] are correct, the quantity of mercury mobilized by burning fossil fuels is as great if not greater than the quantity mobilized and transported by natural waters.

Airborne particulates and gaseous species can be significant sources of trace elements for natural waters when deposited as dry fallout or rainout. Evidence [6,11] indicates the general order of volatility of some trace metals, listed as volatile oxides, sulfates, carbonates, silicates and phosphates, as follows: As, Hg > Cd > Pb > Ag, Zn > Cu > Sn. For trace metals in the elemental state the order is: Hg > As > Cd > Zn > Sb > Ag, Sn, Cu. The sulfides of some trace metals are volatile in the following sequence: As, Hg > Sn > Cd > Sb, Pb > Zn > Cu > Co, Ni, Ag. For example, lead and mercury, in several forms, are released to the atmosphere by volcanic activity, forest fires, smelting operations, the burning of fossil fuels and, for lead, the burning of leaded gasoline. Of these sources, man's activities contribute the major amount [12-15]. The washout of lead during precipitation has been found to be greater than 50 mg/l in rural areas and up to 490 mg/l in urban areas [16]. Recent studies of the concentrations of mercury and lead in Greenland glaciers indicate that concentration levels for recent times are 500 and 2 times greater, respectively, than values predating 800 B.C.

Industrial and Domestic Wastes

Few reliable estimates are available on the quantities of trace metals entering natural waters through industrial and domestic waste effluents. With increasingly strict state and federal effluent standards being imposed to protect receiving waters, more reliable data may be generated through effluent monitoring programs. Water quality studies on the Coeur d'Alene River [17] indicate significant increases in zinc (4.4 ppm max) and cadmium

(0.45 ppm max) in reaches downstream from effluent inputs of industry, mining and domestic sources. Interestingly, Pb, Cu, Ni, and Cr were all below detectable limits. Trace metals such as Pb, Zn, Cd and Cu are commonly found in domestic waste waters in the mg/l range. Leachate from sanitary landfills can have substantial levels of Cu (~0.5 ppm), Zn (~50 ppm), Pb (~0.3 ppm) and Hg (~60 ppb) but flows are generally low [18].

In any case, trace metals from whatever source, once having entered natural water systems, are subject to the complex chemical and biochemical interactions occurring in the natural environment. These interactions undoubtedly involve exchange reactions, adsorption/desorption phenomena, as well as possible redox reactions. The nature of the aqueous metal ions will, to a large degree, influence chemical behavior.

NATURE OF AQUEOUS METAL IONS

Natural waters and their environs are complex heterogeneous electrolyte systems containing both organic and inorganic matter. It is doubtful whether any attempt to describe these complex systems in full detail will ever be complete. However, we can gain useful insights into the chemistry of natural waters by application of equilibrium models to simple homogeneous and heterogeneous systems. To improve agreement between equilibrium models and natural water behavior, nonequilibrium kinetic models need to be developed. While this remains a future goal, we will deal here in some detail with various equilibrium models.

Natural Waters as Electrolyte Solutions

Natural waters are electrolyte solutions in contact with solids of macroscopic to colloidal dimensions. In addition to simple ion-ion interactions, there will also be ion-solvent, ion-solid, and solvent-solid interactions that may be important in the distribution and transport of metal ions between the solid and aqueous phases. Hence, the usual allowance for deviations from ideal behavior due to ion-ion interactions developed by

Debye and Hückel needs to be supplemented in various ways to allow for changes due to the other interactions. A short discussion of ion-solvent interactions in natural electrolyte systems will be useful for understanding subsequent material in this paper.

Solute-Solvent Interactions

In a solution of an electrolyte, it is often necessary to have a detailed knowledge of the species present. New ions or uncharged molecules resulting from interactions in the solution may behave quite differently from the constituent ions of the electrolyte. Some properties (*e.g.*, solubility, adsorption) may be profoundly affected, and the environmental chemist, in order to understand these phenomena in natural systems, will need to know the nature of the species present. There are a number of formidable difficulties in the analysis of such systems. During the past 20 years, a great deal of work has been done on model systems but only recently has work begun on natural aquatic systems. In general, we may regard the elucidation of the structure and composition of a complex electrolyte solution as a difficult problem requiring as many independent lines of attack as possible.

Most natural freshwaters contain concentrations of strong electrolytes sufficient to yield ionic strengths in the general range of 10^{-3} to 10^{-2}. A major characteristic of strong electrolytes (*e.g.*, the alkali halides and the alkaline earth halides) is the general absence of any form of association. In freshwaters, where concentrations are generally small, there should be little association between strong electrolyte species; however, brines and seawater are another matter altogether, and the presence of complexing ligand and chelate species adds to the complexity of the system.

The reason for the ready solubility of so many electrolytes (strong and weak) in water is the high dielectric constant for this solvent, which in turn is due to the polar nature of the water molecule and to the fact that its dimensions favor a tetrahedrally coordinated structure. In a uniform field a dipole experiences only a turning moment, but the field near an ion is highly divergent and therefore nonuniform, and a dipole near an ion is subject to both orienting and attractive forces.

It is obviously of the utmost importance to our understanding of natural electrolyte systems that we should know what the kinetic entity, or ion, is, whether it is the bare ion or whether it carries with it water molecules firmly bound enough to be regarded as part of the ion and if so, how many such molecules there are. (It must be admitted that in general we do not know this with any great certainty.) One major difficulty is that it is not possible to state quite unambiguously what is meant by a water molecule being bound to the ion. As we shall see later in the discussion on interactions at the solid/solution interface, the energy associated with water-ion interactions has implications in terms of the overall behavior of metals in natural aquatic systems.

There are few good examples where the inner sheath of water molecules is tightly bound in the long-term sense, possibly by coordinate bonds. In a series of elegant experiments, Hunt and Taube [19,20] have shown that the ion $[Cr(H_2O)_6]^{3+}$ exchanges water with the bulk solvent very slowly (half-life of 40 hours), while the water molecule of the $[Co(NH_3)_5H_2O]^{3+}$ ion exchanges in about half the time (half-life of 24.5 hours) [21]. Both chromium and cobalt are transition metals with a marked tendency to form coordinate bonds in aqueous solutions. In addition, data for the entropy of solution of many metal ions in water can be explained by assuming a layer of firmly bound water molecules around the ions, with the ion-solvent interaction outside this layer being much weaker [22]. The behavior of the dielectric constant close to the ion is consistent with the above discussion.

Complexity of Natural Aquatic Systems

The solubility of minerals and coordinative compounds of metal ions is perhaps the most important phenomena responsible for the presence of trace metals in natural waters. The important parameters controlling the solubility of mineral phases are the pH of solution, the type and concentration of complexing inorganic and organic ligands and chelating agents, the oxidation state of the mineral components, and the redox environment of the system. How these parameters affect the solubility of heavy metal minerals and the solution compositions will be discussed in more detail.

The electrolyte composition of natural waters varies widely with location. However, the heavy metals are usually dilute both with respect to the total electrolyte composition and on an absolute scale. Hence, the homogeneous solution behavior of simple free metal ions surrounded by their hydration shells, M^{n+}(aq), is generally adequately described by electrolyte theory (the activity coefficients may be estimated from the extended Debye-Hückel theory). On the other hand, rarely does a heavy metal exist in natural waters as a simple ion surrounded by H_2O molecules. Quite commonly the concentrations of anionic species in water, e.g., OH^-, Cl^-, $SO_4^=$, HCO_3^-, organic acids and amino acids, are sufficient to form complexes with the metal ions by replacing coordinated water. Regardless of the strength of the bonding, these complexation reactions are usually quite rapid, with only a few exceptions. For example, metals like Cr(III) with slow aquo exchange react with high OH/metal ratios such that polymers form. These hydroxo polymers condense very slowly with time to yield crystalline oxide or hydroxide phases, e.g., Al(III) [83].

In some instances due to changes in solution composition the solubility of new solid phases may be exceeded and precipitates formed. Depending on the solubility of the new compound, the distribution between solution and solid phases may be altered. For instance, if the ligand is HS^- or $S^=$ the solution concentration of metals will be very small due to the low solubility of the heavy metal sulfides at neutral pH conditions.

Computational techniques are now available to allow calculation of rather complex equilibrium systems [23]. Hence, provided that sufficient information is available concerning the concentration of complexing ligands, the metals present and the thermodynamic data relating them, the distribution of trace metals among aqueous species and solids may be calculated. Even though the computed numbers may be obtained from reliable values for concentrations and stability constants, it is unlikely that the description of solution composition will be accurate because processes such as ion exchange and adsorption have been neglected in the thermodynamic model. When trace metals are at low concentrations these interfacial processes can be expected to be responsible for deviations from the equilibrium models. A more detailed discussion of the influence of interfacial phenomena is given later. Prior to such a discussion it is necessary to review briefly the interactions occurring between metal ions and other solution species.

Hydrolysis Reactions of Heavy Metals

Salts of the heavy metal ions show varying degrees of acidity when dissolved in water. This observation is explained by the reaction of the metal ion with the solvent water, known as hydrolysis, where protons leave the primary or coordinated water molecules in the hydration shell of the metal ion. For example, when aluminum nitrate is dissolved in water one of the reactions that occurs is:

$$Al(H_2O)_6^{3+} + H_2O \xrightleftharpoons{^*K_1} Al(H_2O)_5OH^{2+} + H_3O^+ \quad (1)$$

where $^*K_1 = 10^{-5.0}$.

This reaction may be thought of as dissociation of the coordinated water molecule by the electrical field of the small highly charged ion. An equivalent reaction, not showing the waters of hydration, is

$$Al^{3+} + OH^- \xrightleftharpoons{K_1} AlOH^{2+} \quad (2)$$

where $K_1 = 10^{-9.0}$.

Here the reaction represents the exchange of OH^- for H_2O or the abstraction of H^+ from coordinated water to form a water molecule in solution. The extent to which this simple hydrolysis reaction and others may be measured by an equilibrium constant is:

12 Aqueous-Environmental Chemistry of Metals

$$*K_1 = \frac{[MOH^{(n-1)+}][H^+]}{[M^{n+}]} \qquad (3)$$

which expressed in logarithmic form yields

$$p*K_1 = -\log \frac{[MOH^{(n-1)+}]}{[M^{n+}]} + pH \qquad (4)$$

Hence, when pH = $p*K_1$, the activity ratio $[MOH^{(n-1)+}]/[M^{n+}]$ is unity; consequently this is a useful parameter in discussion of aqueous metal solutions.

Most highly charged metal ions, e.g., Hf^{4+}, Th^{4+}, Fe^{3+}, and Cr^{3+}, are strongly hydrolyzed in aqueous solution and have low $p*K_1$ values, whereas many divalent metals, e.g., Cu^{2+}, Pb^{2+}, Ni^{2+}, Co^{2+}, Zn^{2+}, hydrolyze in the range of pH of natural waters. The alkaline earth metals (e.g., Ca^{2+}) hydrolyze only in basic solutions. The hydrolysis parameter $pK_1 = p*K_1 - pK_w$ varies approximately with the electrostatic energy ratio Z^2/r [24,25] showing the importance of the ionic charge in hydrolytic reactions. Hydrolysis may proceed further by loss of more protons from the coordinated water in a stepwise manner. For example:

$$Fe(H_2O)_5OH^{2+} + H_2O \xrightleftharpoons{*K_2} Fe(H_2O)_4(OH)_2^+ + H_3O^+ \qquad (5)$$

Discussion of mathematic models for these equilibria and graphical representation is given by Butler [26], Sillén [27] and Stumm and Morgan [28]. Each of these species is a simple discrete ionic species in solution. Sillén and co-workers have been prominent in showing that other hydrolytic products containing more than one metal are also formed. For example:

$$2\,FeOH^{2+} \rightleftharpoons Fe_2(OH)_2^{4+} \qquad (6)$$

and

$$2 \text{CuOH}^+ \rightleftharpoons \text{Cu}_2(\text{OH})_2^{2+} \tag{7}$$

These multinuclear or polynuclear complexes vary in kind from small discrete ions to large high molecular weight polymeric compounds, which may generally be considered as kinetic intermediates of insoluble metal oxides often forming very slowly from aqueous solution [29,30]. The slow kinetics of this condensation reaction is one of the reasons that wide variance is often observed in the literature on the values of these hydrolysis constants. This is well demonstrated in articles by Parks [82] and Smith [83] for Al(III) in water. (Further discussion on this point is presented in Chapter 9.)

If only monomeric hydrolysis species are present in a solution, the fraction of the metal as a given hydrolysis product is a function of only the stability constants and the solution pH. This can be shown from the mass balance and algebraic manipulation.

$$[M]_{total} = [M^{n+}] + [MOH^{(n-1)+}] + [M(OH)_2^{(n-2)+}] + \ldots \tag{8}$$

$$\therefore [M]_{total} = [M^{n+}] \left[1 + \frac{{}^*K_1}{[H^+]} + \frac{{}^*K_1 {}^*K_2}{[H^+]^2} + \frac{{}^*K_1 {}^*K_2 {}^*K_3}{[H^+]^3} \ldots \right] \tag{9}$$

$$\therefore a_M = \frac{[M^{n+}]}{[M]_{total}} = \left[1 + \frac{{}^*K_1}{[H^+]} + \frac{{}^*K_1 {}^*K_2}{[H^+]^2} + \frac{{}^*K_1 {}^*K_2 {}^*K_3}{[H^+]^3} \ldots \right]^{-1} \tag{10}$$

Similarly,

$$a_{MOH} = \frac{[M(OH)^{(n-1)+}]}{[M]_{total}} =$$

$$\left[\frac{[H^+]}{*K_1} + 1 + \frac{*K_1 *K_2}{*K_1 [H^+]} + \frac{*K_1 *K_2 *K_3}{*K_1 [H^+]^2} + \ldots \right]^{-1} \quad (11)$$

Thus for monomeric species only the fractional concentration of any species is independent of the total concentration.

During the past twenty years careful research into hydrolysis of aqueous metal ions by Sillén's group at Stockholm [31-33] has demonstrated the importance of polymeric species in accounting for the experimental data, especially in the moderate to high concentration ranges just below saturation with respect to the insoluble hydroxide precipitate. The importance of polymeric species at higher concentrations of metal salt solutions can be seen from the dependence of the fractional concentration of hydrolysis species on the concentration of metal present. For example, consider the case where a metal solution contains a number of monomeric species and at least one polymeric hydrolysis product, e.g., $M_2(OH)_2^{(2n-2)+}$, which may be characterized by the stability constant $*\beta_{22}$, where

$$*\beta_{22} = \frac{[M_2(OH)_2^{(2n-2)+}][H^+]^2}{[M^{n+}]^2} \quad (12)$$

The total mass balance for this solution gives

$$[M]_{total} = [M^{n+}] + [MOH^{(n-1)+}] + 2[M_2(OH)_2^{(2n-2)+}] +$$

$$+ [M(OH)_2^{(n-2)+}] + \ldots \quad (13)$$

and hence

$$a_M = \frac{[M^{n+}]}{[M]_{total}} = \left[1 + \frac{*K_1}{[H^+]} + \frac{2[M^{n+}]*\beta_{22}}{[H^+]^2} + \frac{*K_1*K_2}{[H^+]^2} + \cdots\right]^{-1} \tag{14}$$

$$a_{MOH} = \frac{MOH^{(n-1)+}}{[M]_{total}} = \left[\frac{[H^+]}{*K_1} + 1 + \frac{2[M^{n+}]*\beta_{22}}{[H^+]*K_1} + \frac{*K_1*K_2}{*K_1[H^+]} + \cdots\right]^{-1} \tag{15}$$

and

$$a_{M_2(OH)_2} = \frac{[M_2(OH)_2^{(2n-2)+}]}{[M]_{total}}$$

$$= \left[\frac{[H^+]^2}{*\beta_{22}[M^{n+}]} + \frac{*K_1[H^+]}{*\beta_{22}[M^{n+}]} + 2 + \frac{*K_1*K_2}{*\beta_{22}[M^{n+}]} + \cdots\right]^{-1} \tag{16}$$

Thus as the concentration $[M^{n+}]$ is increased, the formation of polymeric species is favored.

The selection of the stability constants for particular polynuclear complexes that best fit the experimental data is a matter of some complexity. Special computation methods have been developed by Sillén's group and others. A large and comprehensive compilation of stability constants for inorganic and organic ligands bound to metal ions was published in 1958; this has since been revised and extended by Sillén and Martell [34,35] in 1964 and 1971. In the section concerned with hydrolytic equilibria of aqueous metal ions there is quite often a large variation in the reported values of a particular complex even though it would appear that the experimental conditions such as temperature, pressure, ionic strength, anion type and the presence of solid phases are similar for the determinations. This occurs because the solubility of ions in equilibrium with solid phases depends on the particle size of the solid and also because the solid phases may be difficult to identify.

Complexation of Metal Ions in Solution

The chemical composition of natural waters is derived from weathering products, atmospheric fallout and rainout as well as from agricultural, industrial and domestic wastes and effluents. Consequently, there are a number of natural and synthetic organic and inorganic complexing agents available to react with aqueous metal ions in addition to the solvent water. Among the more important inorganic complexing ligands are Cl^-, $SO_4^=$, HCO_3^-, F^-, and sulfide and phosphate species.

In much the same manner that reactions between metal ions and water yield various soluble and insoluble species, a similar suite of soluble complex ions and insoluble phases may result from reactions with inorganic anions, depending on the metal concentration, ligand concentration, and pH. Tables 1.1, 1.2 and 1.3 present typical kinds of interactions for several trace metals [34-37]. For example, mercuric ion, Hg^{2+}, reacts with H_2S in acid sulfide solutions to precipitate HgS. However, if the solution conditions are adjusted to high pH and high HS^- or $S^=$ concentrations, dissolution occurs due to the formation of soluble complexes:

$$HgS(s) + HS^- = HgS_2H^-, \qquad pK = 5.28 \qquad (17)$$

and

$$HgS(s) + S^= = HgS_2^=, \qquad pK = -0.57 \qquad (18)$$

Hence, the formation of insoluble ligand complexes can greatly affect the distribution of metals between the solid and

Table 1.1

Solubility and Complex Formation Equilibria for Mercury

No.	Equilibria	log K (25°C)
	I. Solubility Equilibria	
1.	$HgO(s) + H_2O = Hg(OH)_2 \text{ (aq)}$	-3.7
2.	$HgO(s) + H_2O = Hg^{2+} + 2OH^-$	-25.7
3.	$Hg(l) = Hg(aq)$	-6.50
4.	$HgCl_2(s) = Hg^{2+} + 2Cl^-$	-13.8
5.	$Hg_2Cl_2(s) = Hg_2^{2+} + 2Cl^-$	-18.00
6.	$Hg_4OCl_2(s) + H_2O = 2Hg_2^{2+} + 2Cl^- + 2OH^-$	—
7.	$2Hg_2OCl(s) + 2H_2O = 2Hg^{2+} + Hg_2^{2+} + 2Cl^- + 4OH^-$	—
8.	$HgSO_4 \cdot 2HgO(s) + 2H_2O = 3Hg^{2+} + SO_4^{=} + 4OH^-$	—
	II. Complex Formation Equilibria	
9.	$Hg^{2+} + H_2O = HgOH^+ + H^+$	-3.4
10.	$Hg^{2+} + 2H_2O = Hg(OH)_2^{\circ} + 2H^+$	-6.00
11.	$Hg^{2+} + 3H_2O = Hg(OH)_3^- + 3H^+$	-20.7
12.	$Hg^{2+} + Cl^- = HgCl^+$	6.7
13.	$Hg^{2+} + 2Cl^- = HgCl_2^{\circ}$	13.2
14.	$Hg^{2+} + 3Cl^- = HgCl_3^-$	14.2
15.	$Hg^{2+} + 4Cl^- = HgCl_4^{=}$	15.2
16.	$HgCl_2^{\circ} + H_2O = HgOHCl^{\circ} + Cl^- + H^+$	-9.6
17.	$HgCl_2 + 2H_2O = Hg(OH)_2^{\circ} + 2Cl^- + 2H^+$	-19.6
18.	$Hg^{2+} + Cys = HgCys^{2+}$	43.57*
19.	$Hg^{2+} + Gly^- = (HgGly)^+$	10.3**
20.	$(HgGly)^+ + Gly^- = Hg(Gly)_2^{\circ}$	8.9

*Cys = cysteine   **Gly = glycine

Table 1.2

Solubility and Complex Formation Equilibria for Lead, Zinc and Cadmium

No.	Equilibria	log K (25°C)
	I. Solubility Equilibria	
1.	$PbCO_3(s) = Pb^{2+} + CO_3^=$	-13.00
2.	$Pb(OH)_2(s) = Pb^{2+} + 2OH^-$	-14.93
3.	$PbSO_4(s) = Pb^{2+} + SO_4^=$	- 7.89
4.	$Pb_3(OH)_2(CO_3)_2(s) = 3Pb^{2+} + 2OH^- + 2CO_3^=$	-18.80
5.	$ZnCO_3(s) = Zn^{2+} + CO_3^=$	-10.00
6.	$Zn(OH)_2(s) = Zn^{2+} + 2OH^-$	-15.52
7.	$CdCO_3(s) = Cd^{2+} + CO_3^=$	-13.74
8.	$Cd(OH)_2(s) = Cd^{2+} + 2OH^-$	-13.79
	II. Complex Formation Equilibria	
9.	$Pb^{2+} + OH^- = PbOH^+$	5.85
10.	$Pb^{2+} + 2OH^- = Pb(OH)_2^\circ$	10.80
11.	$Pb^{2+} + 3OH^- = Pb(OH)_3^-$	13.92
12.	$Pb^{2+} + Cl^- = PbCl^+$	1.62
13.	$Pb^{2+} + 2Cl^- = PbCl_2^\circ$	1.83
14.	$Zn^{2+} + OH^- = ZnOH^+$	4.95
15.	$Zn^{2+} + 2OH^- = Zn(OH)_2^\circ$	12.89
16.	$Zn^{2+} + 3OH^- = Zn(OH)_3^-$	14.22
17.	$Zn^{2+} + 4OH^- = Zn(OH)_4^=$	15.48
18.	$Zn^{2+} + Cl^- = ZnCl^+$	- 0.56
19.	$Zn^{2+} + SO_4^= = ZnSO_4^\circ$	2.8
20.	$Cd^{2+} + OH^- = CdOH^+$	4.59
21.	$Cd^{2+} + 2OH^- = Cd(OH)_2^\circ$	8.93
22.	$Cd^{2+} + 3OH^- = Cd(OH)_3^-$	9.58
23.	$Cd^{2+} + Cl^- = CdCl^+$	2.08
24.	$Cd^{2+} + OH^- + Cl^- = CdOHCl^\circ$	5.87
25.	$Cd^{2+} + SO_4^= = CdSO_4^\circ$	2.76

Table 1.3

Sulfide Solubility Equilibria for Heavy Metals at 25°C

No.	Equilibria	log K (25°C)
1.	$MnS(s) = Mn^{2+} + S^=$ (crystaline)	12.85
2.	$MnS(s) = Mn^{2+} + S^=$ (precipitated)	15.7
3.	$FeS(s) = Fe^{2+} + S^=$ (pyrite)	18.75
4.	$FeS_2(s) = Fe^{2+} + S_2^=$	26.1
5.	$CoS(s) = Co^{2+} + S^=$	21.3
6.	$ZnS(s) = Zn^{2+} + S^=$ (precipitated)	22.05
7.	$ZnS(s) = Zn^{2+} + S^=$ (wurtzite)	22.80
8.	$ZnS(s) = Zn^{2+} + S^=$ (sphalerite)	25.15
9.	$\alpha\text{-}NiS(s) = Ni^{2+} + S^=$	20.8
10.	$\gamma\text{-}NiS(s) = Ni^{2+} + S^=$	27.75
11.	$CdS(s) = Cd^{2+} + S^=$	27.2
12.	$PbS(s) = Pb^{2+} + S^=$	28.2
13.	$CuS(s) = Cu^{2+} + S^=$	36.2
14.	$HgS(s) = Hg^{2+} + S^=$ (metacinnabar)	52.2
15.	$HgS(s) = Hg^{2+} + S^=$ (cinnabar)	53.6

aqueous phases by the formation of metal-ligand precipitates. This is particularly true at the sediment-water interface of natural systems where composition and environmental gradients are usually the largest.

Some ligands that form soluble complex ions and molecules can also react in another way by dissolving other insoluble phases. For example, mercuric ions form very stable complexes with chloride ions (*e.g.*, $HgCl^+$, $HgCl_2^°$, $HgCl_3^-$, and $HgCl_4^=$) while the $HgCl_2(s)$ phase remains quite soluble. In chloride-free solutions mercuric ions hydrolyze at low pH (~3) and may form a solid precipitate, $HgO(s)$, if the concentration is greater than ~50 ppm (~2.5 x 10^{-4} mol/l). However, in the presence of excess chloride ion, say 10^{-3} M, the Hg^{2+} reacts to form $HgCl_2^°$, thus decreasing free Hg^{2+} so that hydrolysis products only become dominant at higher pH values (pH 6~7). Therefore, complexation by one kind of ligand ion results in the increased metal ion solubility of the precipitate of another ligand.

Other possibilities may also occur. For example, some of the heavy metals may form "double salts," that is, the coordination sheath may contain both OH^- and other ligands, *e.g.*, Cl^- or $SO_4^=$ (reactions 6, 7, and 8 in Table 1.1 and reaction 4 in Table 1.2). These products again show different solubilities that can affect the solid/solution distribution of the metal ions. An example of a complex metal ion-ligand system is the calcium phosphate sequence of solids where there is competition between phosphate, carbonate, and OH^- for coordination with calcium [38]. The resulting double salt solid phase, hydroxyapatite, shows solubility and surface chemical characteristics very different from either the simple calcium phosphate or calcium carbonate solid phases [142-145].

Most inorganic ligands of importance in natural waters tend to be present at orders of magnitude higher concentration than the trace metal ions they tend to complex. Each metal ion has a speciation in simple aqueous solutions that is dependent upon the stability of the hydrolysis products and the tendency of the metal ion to form complexes with other inorganic ligands. For example, Pb(II), Zn(II), Cd(II), and Hg(II) each form a series of complex species (Tables 1.1 and 1.2) when in the presence of Cl^- and/or $SO_4^=$ at concentrations similar to seawater. The pH at which hydrolysis products are formed in any significant percentage is dependent upon the concentration of the ligand competing with OH⁻ for the metal ion. For example, as shown in Figure 1.2, equal concentrations of $Hg(OH)_2^0$ and $HgCl_2^0$ are present at pH ~ 6.8 for $pCl_T = 3$ and at pH ~ 8.8 for $pCl_T = 1$, indicating significant competition for Hg(II).

Chelation with Organic Ligands

A large variety of natural and synthetic organic matter is now present in natural waters. This includes the natural degradation products of plant and animal tissue, *e.g.*, amino acids and humic acids [39,40], and organic species derived from chemicals applied either deliberately or inadvertently by man. Examples of the latter case are detergents and cleaning agents, NTA, EDTA, pesticides, ionic and nonionic surfactants with various functional groups, and a large group of synthetic macrocyclic compounds [39-41].

The term "metal-organic" is used here to describe structural configurations in which the metal is bonded to organic matter by way of (1) carbon atoms yielding organo-metallic compounds, (2) carboxylic groups producing salts of organic acids, (3) electron-donating atoms, O, N, S, P, etc., forming coordination complexes, or (4) π-electron-donating arrangements (olefinic bonds, aromatic rings, etc.). Our main interest here is in coordination compounds that are preserved, formed, or destroyed during processes of sedimentation.

Apart from a few metallo-porphyrin enzymes, the binding of metals in proteins is little known. Porphyrin-type materials are wide spread in living systems. Porphyrin complexes isolated from various organisms include chlorophylls, iron porphyrins, cyanocobalamin (vitamin B_{12}), and many other compounds in which the tetrapyrrole system is bonded to copper, zinc, cadmium and manganese.

The significance of organics in determining the speciation of trace metal ions has been a subject of continued controversy between investigators of fresh and marine waters. One view is that the high concentrations of Ca(II) and Mg(II) present in most aquatic systems make competition for the low concentrations

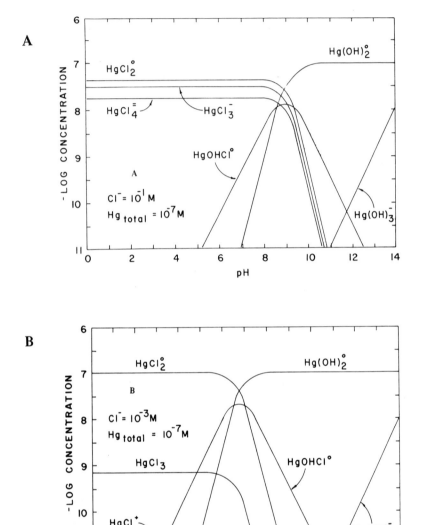

Figure 1.2. Log concentration-pH diagram for mercury at $25°C$. $10^{-7}\ M$ total mercury. **A.** $10^{-1}\ M$ total chloride. **B.** $10^{-3}\ M$ total chloride.

of dissolved organics difficult for trace metals [28,42,43]. From another point of view, numerous analyses have indicated that a large fraction of the metals found in natural waters are in what is called the organic fraction [44-49].

The concentrations of aqueous species in seawater and freshwater are not necessarily the result of equilibrium processes. It may be useful at this point to indicate the difference between thermodynamic stability and lability. Species which are thermodynamically unstable but which, due to kinetic constraints, exist in significant amounts are called nonlabile or inert [50]. Because of these nonequilibrium processes, extrapolation of many thermodynamic stabilities as determined in well-controlled laboratory environments with known solution species to natural conditions requires considerable caution. Although the "apparent" stability of metal ions and natural organics extracted from soils and natural waters has been investigated in a few cases [51-54], the extent of the interaction of these organic compounds with metal ions in the real environment is unknown.

The influence of natural organic chelates and solutes on the adsorption behavior of trace metal ions on natural colloidal materials is not at all understood. The influence of model compounds on the surface reactions of metal ions has been evaluated by only a few investigators and will be discussed in a later section.

Oxidation-Reduction Reactions

Natural waters are, in general, highly dynamic with respect to oxidation-reduction reactions. Typically having great variations in redox environments, most natural waters are far from an equilibrium state. For example, in an estuary or freshwater lake there is a marked difference in redox environment between the surface in contact with oxygen in the atmosphere and the bottom waters at the sediment-water interface. The region in between may also reflect large redox gradients depending upon mixing, diffusion and the extent of biological activity. Exact definition of redox environments in natural water systems is probably not as important as understanding and defining the redox gradients and transport mechanisms acting within the system. In view of the above, it can be said that equilibrium redox calculations are useful to the extent that they facilitate understanding of the redox patterns observed or anticipated in natural water systems. In the absence of kinetic information, equilibrium models provide, at a minimum, boundary conditions within which questions may be framed by comparing observed conditions with the models.

Differences between computed equilibrium boundary conditions and available data on real systems allows valuable insight and speculation into the reasons for the differences, insufficient understanding of chemical reactions or kinetic factors notwithstanding. The major value of any phenomenological model lies basically in the comparative format that allows questions to be asked; for example, one may speculate on the type of dissolved species, solid phases or controlling mechanisms that may be expected under various oxidizing or reducing conditions [55-58].

Because meaningful direct redox measurements are generally not possible in natural waters [55,56], most estimates of the redox environment are indirect and semiquantitative. In any case, because of the highly complex nature of natural aquatic systems, it is quite likely that many redox environments (micro and macro) co-exist in the same system and within a given region, or area redox couples may be present in varying degrees of completeness depending upon kinetic factors. Great care must be exercised both in terms of measurement and interpretation of data.

Changing redox environments can affect trace metals in aquatic systems in two ways: (1) by direct changes in the oxidation state of the metal ion, and/or (2) by redox changes in available and competing ligands or chelates. Typical redox environments found in nature are presented in Figure 1.3. For purposes of discussion, equilibrium pE-pH diagrams are presented for lead and mercury in Figures 1.4, 1.5, and 1.6. The solid phases presented for lead are the most likely phases to be found, thermodynamically, in an aqueous environment. Native lead is not considered since it is stable only at very low pE and high pH. Interestingly, Pb(II) is the only stable oxidation state for lead in most aqueous environments and, hence, the effect of changes in pE or pH conditions affects the combining ligand

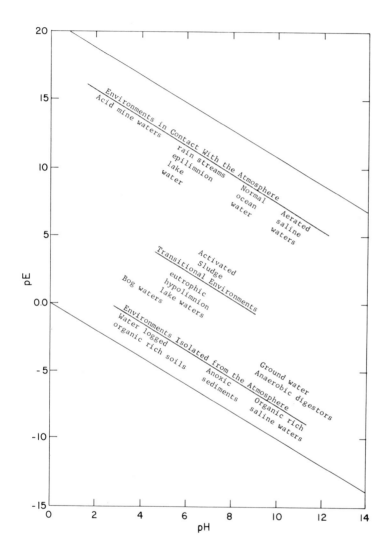

Figure 1.3. Relative positions of various environments as characterized by pE and pH.

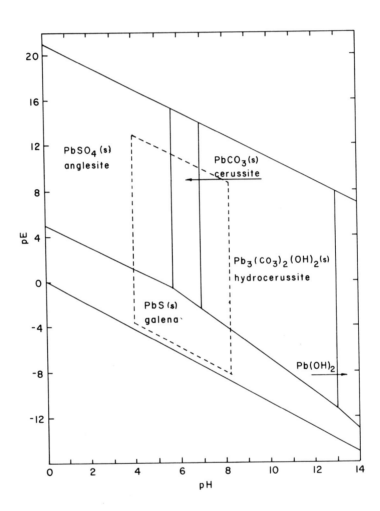

Figure 1.4. Stability fields of Pb solid phases as a function of pE and pH at 25°C and 1 atm total pressure. Diagram calculated for conditions of of 10^{-3} M total carbonate species and total sulfur species in aqueous solution. Dashed lines represent pE-pH conditions common in nature.

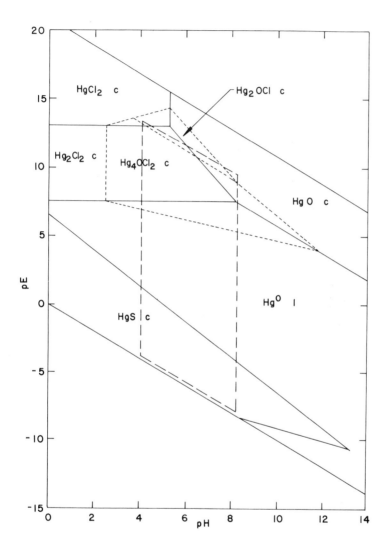

Figure 1.5. Stability fields of Hg solid phases as a function of pE and pH at 25°C and 1 atm total pressure. Diagram calculated for conditions of 10^{-3} M Cl^- and $SO_4^=$ in aqueous solution. Dotted lines represent hypothetical stability fields of eglestonite (Hg_4OCl_2) and terlinguaite (Hg_2OCl). Dashed lines represent pE-pH conditions common in nature.

28 *Aqueous-Environmental Chemistry of Metals*

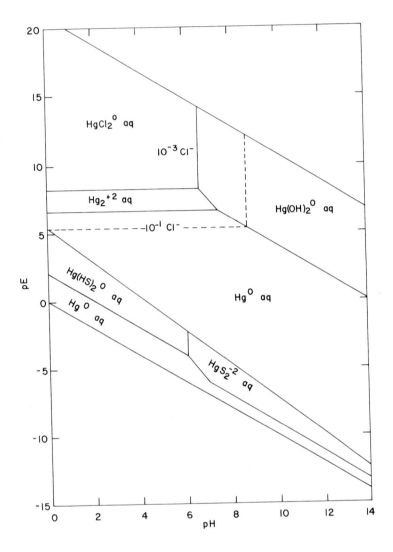

Figure 1.6. Stability fields of Hg aqueous species as a function of pE and pH at 25°C and 1 atm total pressure. Solution contains 10^{-3} M $SO_4^=$ and 10^{-3} or 10^{-1} M Cl^-. Dashed line represents expanded field boundary of $HgCl_2^{\,\circ}$. High solubilities exist over the upper one-third and extreme lower right of the diagram.

rather than the metal itself. By contrast, mercury can easily change oxidation states within the range of natural environmental redox conditions (Figures 1.5 and 1.6) and is, therefore, subject to both direct and indirect influence of changing redox conditions. Mercury is somewhat unusual in that it may be present in different oxidation states in both the condensed phases and in the dissolved aqueous phase. In addition, liquid mercury has a rather high vapor pressure, thus forcing us to consider the possibility of significant transport through the atmosphere.

Interactions in Heterogeneous Sulfide Systems

The sediment-water interface is the site of important interactions controlling and influencing the behavior of trace metals in aquatic systems. One of the major processes occurring in many natural systems at, or just below, the sediment-water interface is the oxidation and reduction of many organic and inorganic compounds. Intense redox activity occurs at the sediment-water interface mainly because of the deposition and accumulation of organic matter and the necessity of molecular oxygen to diffuse down into the sediment interstitial waters, thus creating a large redox gradient.

It is a well-known fact that high organic sediments generally contain large quantities of reduced material, especially sulfides. Since most heavy metal sulfides tend to be rather insoluble (Table 1.3) and since the Fe(II)/Fe(III) redox reaction [59] readily occurs within the redox gradient found near sediment-water interfaces, it is clear that interactions in heterogeneous sulfide systems can be important processes whereby trace metals are retained or released from the aqueous phase.

The major sulfide phases found in natural sediments are those of the iron system [60,61]. Since the solubility of most other heavy metal sulfides is lower than those of the iron sulfide system, it is probable that the oxidative dissolution and reductive precipitation of sedimentary iron sulfides are important processes whereby trace metals are released or retained in some sediments.

A somewhat different sulfide system may be useful in depicting the complex processes and mechanisms occurring in reduced sulfide systems. For example, in a heterogeneous system containing an insoluble precipitate (ZnS) and an aqueous metal ion that forms a precipitate of even lower solubility [Cu(II)(aq) which may form CuS], there is the possibility of both adsorption and lattice exchange reactions. However, the energetics of the lattice exchange reactions are usually much greater than adsorption from aqueous solution, so that at equilibrium the main control on the second metal will be solubility.

The nature of the reaction between Cu^{2+} (and also Ag^+) and ZnS in slightly acidic solutions has been shown by Gaudin, Fuerstenau and co-authors [62-64] to be an exchange reaction in which a more insoluble sulfide forms a coating on the sphalerite surface and releases Zn^{2+} to the aqueous phase. For example,

$$Cu^{2+} + ZnS \rightleftharpoons Zn^{2+} + CuS \qquad (19)$$

and

$$2\,Ag^+ + ZnS \rightleftharpoons Zn^{2+} + Ag_2S \qquad (20)$$

where the driving force for the reaction is essentially the difference in the solubility of the sulfides.

This result can be generalized for various pH values in a log concentration-pH diagram for the aqueous Zn(II)-Cu(II)-sulfide system for reducing or low pE conditions such as in Figure 1.7A. In this diagram it was assumed that there was a total of 97 mg/l of ZnS and 6.4 mg/l ion Cu(II) and that the volume of the vapor phase was negligible so that loss of H_2S need not be considered. Details of mass balances and stability constants used are given by James and Parks [146].

The diagram shows that at equilibrium a CuS coating will exist on ZnS from low pH, < 1, to high pH, > 11. In slightly basic solutions ZnO may also precipitate, but this depends on the amount of Zn^{2+} released from the ZnS crystal by exchange with Cu^{2+}. The ratio $[Cu^{2+}]/[Zn^{2+}] = 10^{-12}$ is a constant when both CuS and ZnS phases are present. However, after a few surface layers of CuS have been formed on the ZnS crystal in the real nonequilibrium process, Zn^{2+} and $S^=$ cannot diffuse out of the crystal fast enough to remove the Cu^{2+} to the equilibrium level. Hence, the ratio may be much greater than 10^{-12}. For example, Gaudin, Fuerstenau and Mao [62] have reported that

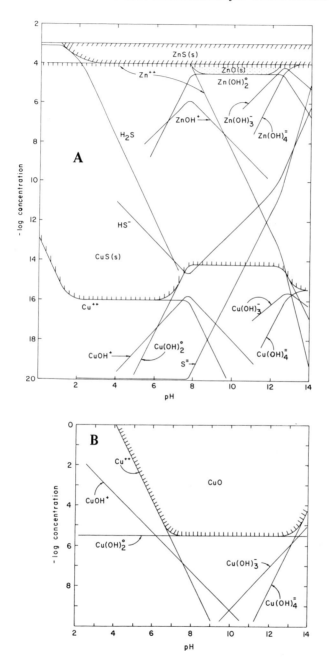

Figure 1.7. Log concentration-pH solubility diagrams. **A.** Heterogeneous ZnS-Cu(II)-H$_2$O system at 25°C. **B.** CuO-H$_2$O system at 25°C.

the uptake of Cu^{2+} on ZnS of approximately 0.8×10^{-6} mol/gm was constant over a residual or "equilibrium" solution concentration range from 10^{-5} to 10^{-4} mol/l at 30 minutes after its addition at approximately pH 6. Kinetic studies showed only very slow exchange after 30 minutes reaction. That is, the "concentration" of all the aqueous Cu(II) species is greater than in the equilibrium diagram, Figure 1.7A. This means that the concentration of Cu(II) may well be sufficient to allow nucleation and precipitation of $Cu(OH)_2$ in neutral solutions to form surface Cu(II) hydroxide coatings on the sulfide mineral [69] as indicated in Figure 1.7B.

Also, because only small amounts of Zn^{2+} species are exchanged with the Cu^{2+} it is unlikely that ZnO will form as indicated in the figure. The Zn^{2+} displaced from ZnS by Cu^{2+} reported by Gaudin, et al. [62] is equivalent to 5×10^{-5} mol/l. The log concentration – pH diagram, Figure 1.7A – indicates that ZnO would not precipitate. At low pH the solubility product of CuO or $Cu(OH)_2$ is not exceeded and at high pH the CuO begins to dissolve to form anionic complexes and expose the CuS surface as shown by Figures 1.7A and 1.7B.

In this kind of system the interaction between the aqueous metal and the sulfide mineral is primarily an exchange reaction with the crystal. This may be accompanied by precipitation of metal hydroxy complexes in the neutral pH region. This is unlike the adsorption on oxide minerals where the reaction may be either ion exchange, specific adsorption or precipitation of hydroxides. Adsorption in the Cu(II) reaction with ZnS is only the initial stage in the reaction, and the energetics of the adsorption step are overwhelmed by the exchange free energy.

If one wishes to consider only pure adsorption processes on sulfide minerals, one must use a relatively soluble aqueous metal ion and a relatively insoluble sulfide solid phase. The obvious sulfides to choose in this respect are Ag_2S and HgS. A great deal of work has already been done on the solubility and the electrical double layer structure [66,67] of the Ag_2S-H_2O system; hence, this is probably the most promising sulfide to work with in future adsorption studies.

HgS is the most insoluble sulfide in acid and neutral pH [68] and also provides a convenient solid sulfide phase. Studies currently in progress on the adsorption of Zn(II) on HgS indicated that Zn(II) adsorbs on HgS in the same manner as on SiO_2 and Al_2O_3 [69,70].

A more detailed discussion of the role of adsorption processes follows in the next section. It is clear, however, that in many instances multiple processes may be acting in concert to control the concentration and behavior of trace metals in aquatic systems. The combination of significant redox gradients, adsorption/desorption phenomena and precipitation or coprecipitation/dissolution make for difficult interpretation of field information and, therefore, necessitate a continued effort to gain greater insight into such complex natural systems.

INTERACTIONS WITH SOLID/SOLUTION INTERFACES

When solutions of trace metals or other dilute solutes come in contact with solid phases, the concentration of the metal (or solute) usually decreases through its association with the solid phase. The uptake of the aqueous metal ions has been attributed to a number of processes, for example, adsorption, ion exchange, or coprecipitation, for which quantitative and semiquantitative models have been proposed. These processes may be responsible for reducing the concentration of dissolved metals in natural waters below the concentration prescribed by equilibrium solubility models. Through association of metals with particulate matter the distribution pattern of metals in the aqueous environment will be affected by the chemical and physical conditions of the system. The emphasis of this discussion is placed on the chemical properties of the interfacial systems.

Nature of the Solid/Solution Interface

Inorganic Systems

Surface charge exists along the fracture and cleavage surfaces of most minerals due to the rupture of bonds. When such minerals are brought in contact with water, these surface charges respond by adsorbing H_2O molecules, which can dissociate. Dissolution occurs until equilibrium is established. This may be represented by

$$\underline{MX}^- + H_2O \underset{}{\overset{{}^*K_1}{\rightleftharpoons}} \underline{MXH}^\circ + OH^- \qquad (21)$$

and

$$\underline{MX}H° + H_2O \overset{*K_2}{\rightleftharpoons} \underline{MX}H_2^+ + OH^- \qquad (22)$$

where M_xX_m may be an insoluble metal oxide, sulfide, phosphate, etc., and $\underline{MX}^-$, $\underline{MX}H°$ and $\underline{MX}H_2^+$ represent negative, neutral and positive surface sites. In ion exchange terminology, $\underline{MX}H°$ is an exchanger in the protonated form. A more detailed discussion of alternative charging mechanisms of mineral/solution interfaces and the development of the electrical double layer is given by Parks [24,71] and Healy, et al. [72]. Clay minerals may acquire surface charge by isomorphic structural substitution in the lattice as well as by other mechanisms.

The overall surface charging reaction for minerals involving water can be represented by

$$\underline{MX}^- + 2H^+ \overset{K_s}{\rightleftharpoons} \underline{MX}H_2^+ \qquad (23)$$

where the activity of the negative and positive sites are related to the solution pH

$$pK_s = -\log \frac{\underline{MX}H_2^+}{\underline{MX}^-} - 2\,pH \qquad (24)$$

Hence ½ pK_s is the pH at which the activities of the charged sites are equal and, if the activity coefficients are the same, $\gamma_+ = \gamma_-$, the surface will have net zero charge. Often ½ pK_s will show some agreement with the pH of minimum solubility of the solid phase.

The formation of surface charge in water by either the adsorption of H^+ and OH^- or the preferential dissolution of ions from the surface can be observed by titration techniques to measure the adsorbed charge directly or by electrokinetic measurements to observe interfacial potentials.

The determination of the intrinsic acidity constants, $*K_1$ and $*K_2$, is not as simple as the determination of K_s because as the surface charge builds up, the opposite charges are more and more difficult to remove, i.e., $*K_1$ and $*K_2$ appear to vary.

This is similar to the titration of proteins. However, methods have been developed [73-76] to handle this problem. By making some assumptions about the adsorption of ionic species an approximate relationship between surface charge and pH can be obtained.

Organic Colloids in Natural Waters

Apart from the obvious differences in bulk composition, organic colloids in water show some remarkable similarities to colloidal minerals with respect to interfacial and sorption properties. High molecular weight organics such as humic acids are generally of low solubility and have dissociable hydroxyl and carboxyl functional groups. Surface charge on macromolecular or particulate proteins develops through dissociation or hydrolysis of functional groups, which are components of the amino acids comprising the protein. These functional groups include hydroxyl, carboxyl, sulfhydryl, amino and imino groups [76]. Hence, depending upon the solution pH, the surface may be charged or neutral. Organic colloids may come from vegetable or animal decay matter and, depending upon the source material, may have varying proportions of the different functional groups. All these types of functional groups may either generate a surface charge that governs electrostatic adsorption or form direct organic-metal bonds [50,52,77,78].

In addition to the importance of purely organic colloids in natural interfacial systems, both high molecular weight macromolecules and smaller organic compounds can adsorb on and modify inorganic surfaces [79]. For example, surfactants adsorbing at the air-water interface and polymers forming coatings on suspended matter may be partially responsible for the concentration of metal ions in surface waters and on suspended matter [49,80].

It is apparent from this brief discussion that natural interfacial systems are complex when taken in total. The surface state and reactions of minerals in natural waters have been alluded to and a discussion of the interactions that occur when metal ions are present follows.

Adsorption Behavior of Trace Metals

Even though almost all natural waters contain some amount of organic materials, most experimental work on the adsorptive behavior of trace metal ions has been done in the absence of organic ligands or organic colloids. The discussion here will focus primarily on inorganic systems since more theoretical and experimental information is available for the development of phenomenological models.

The adsorption behavior of hydrolyzable metal ions at the solid/water interface is strongly pH dependent and is characterized by a general agreement between the hydrolysis of the aqueous metal ions and their enhanced adsorption, charge reversal and coagulant properties. Models used to describe the phenomena, including the ion exchange and the specific adsorption of certain species, are compared to another model in which all the species of a hydrolyzable metal ion are considered as potential adsorbates. In this latter model only the lower charged, less strongly hydrated hydrolysis complexes have favorable free energies for the adsorption process.

The current awareness of heavy metals as pollutants in aquatic systems has aroused much interest in the factors that appear to be involved in the distribution of metal ions between sediments and the aqueous phase. The processes described here are of direct application to inorganic sediment systems. Of course, the chemistry of sediments and natural waters is complicated by the presence of a variety of organic solutes, colloids and solid phases. However, the interaction of hydrolyzable metal ions at oxide/water interfaces is a convenient reference system for comparison with other data and studies alluding to adsorption mechanisms.

From a wide range of investigations in different disciplines, it has been shown that adsorption or uptake of aqueous metals by mineral/water surfaces increases abruptly in the pH range where hydrolysis products become a significant fraction of the concentration of the aqueous metal ion. The general interfacial behavior of hydrolyzable metal ions can be summarized as follows:

1. For each metal there is a critical pH range, often less than 1 unit wide, over which the fractional amount of metal adsorbed increases from almost zero to unity, *i.e.*, almost complete adsorption or removal from solution. There appear to be some subtle differences for different substrates, but this usually has only a small effect on the critical pH range. For example, the adsorption density of Fe(III) and Pb(II) on quartz shown in Figure 1.8 increases abruptly when the following types of hydrolysis reactions occur for both iron(III) and lead(II) [81]:

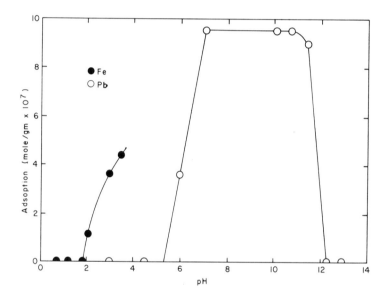

Figure 1.8. Adsorption of Fe(III) and Pb(II) aqueous species on quartz as a function of pH (after Fuerstenau, *et al.* [81]).

Mononuclear species $*K_1$

$$M^{n+} + H_2O \rightleftharpoons MOH^{(n-1)+} + H^+ \quad (25)$$

Polynuclear and high polymeric species

$$2MOH^{(n-1)+} \rightleftharpoons M_2(OH)_2^{(2n-2)+} \rightleftharpoons \text{high polymers} \quad (26)$$

Insoluble precipitate formation

$$2M^{n+} + nH_2O = M_2O_n + 2nH^+ \qquad (27)$$

The usual sequence in the hydrolysis reactions is monomeric species forming polymeric species, which yield crystalline oxide precipitates after long aging times [82-84]. The adsorption of aqueous Co(II), Mg(II) and Ca(II) on quartz [85] also increases abruptly when the pH approaches hydrolysis conditions as shown in Figure 1.9.

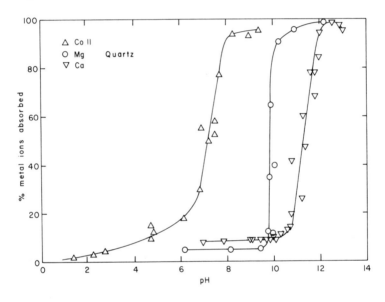

Figure 1.9. Adsorption of metal cations on quartz as a function of pH. Co(II) concentration was initially 1×10^{-4} M. Mg(II) and Ca(II) concentrations were initially 4×10^{-3} M and 3×10^{-3} M, respectively (after Fuerstenau [85]).

2. If the substrate has a negative surface charge and, hence, a negative surface potential, the experimentally observed double layer potential, ζ, usually reverses sign when the metal ion hydrolyzes and metal hydroxides are precipitated. The free aquo species appears to have relatively little effect on the ζ potential of the solid.

Observation of the electrical properties of the mineral-water interface during an adsorption experiment reveals reversal of the sign of the surface charge from - to + when the pH is suitable for high adsorption densities and hydrolysis. This result is illustrated by the data of Mackenzie [86,87] and Fuerstenau, et al. [81] on the electrokinetic behavior of quartz in Fe(III) [86], Co(II) [87], and Pb(II) [81] solutions shown in Figures 1.10 and 1.11. This type of evidence has been interpreted by Healy, et al. [90] to show that the original substrate oxide becomes coated with hydroxides and hydrous oxides of the adsorbing metal ion. The charge reversal at high pH often corresponds to the isoelectric point of the pure oxide of the adsorbing metal ion.

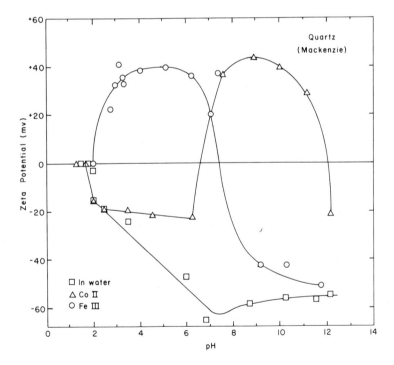

Figure 1.10. Zeta potential-pH relationship for quartz in the presence of 4×10^{-3} M $CO(NO_3)_2$ and 6×10^{-3} M $FeCl_3$.

40 *Aqueous-Environmental Chemistry of Metals*

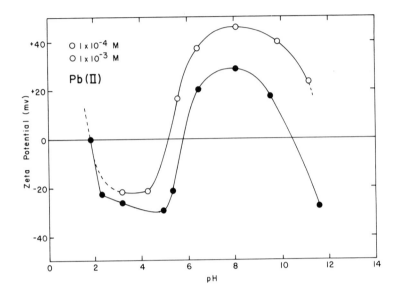

Figure 1.11. Zeta potential-pH relationship for quartz in the presence of 1×10^{-3} M and 1×10^{-4} M Pb(NO$_3$)$_2$.

3. Concurrent with changes in the ζ potential, changes are observed in the stability of colloid dispersions. Generally, rapid coagulation and settling of colloids occurs when the ζ potential decreases to a low magnitude. In some instances, the formation of hydrolysis products is claimed to have been detected by coagulation techniques [91]. The relation between the electrokinetic potential of SiO$_2$ dispersions in aqueous Co(II) solutions with the stability of the dispersion is shown in Figure 1.12. At similar dispersion concentration and metal concentration the approach to zero potential corresponds to rapid coagulation rates, indicated by a low turbidity index.

The importance of metal ion hydrolysis and precipitation in adsorption, electrokinetic and coagulation behavior (and in water treatment processes) is stressed in Figure 1.13 which shows the equilibrium concentration of some iron(III) species as a function of pH [92]. The range of concentration and pH used

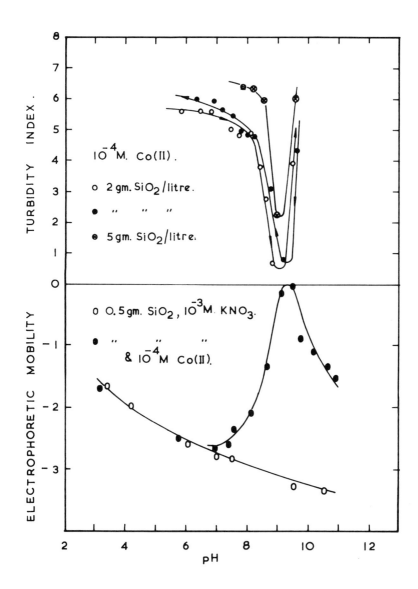

Figure 1.12. Coagulation (turbidity index, arbitrary units) and electrophoretic mobility ($\mu m\ sec^{-1}/volt\ cm^{-1}$) of SiO_2 as a function of pH. 10 m^2/l and 2.5 m^2/l surface area concentration at 1 × 10^{-4} M Co(II).

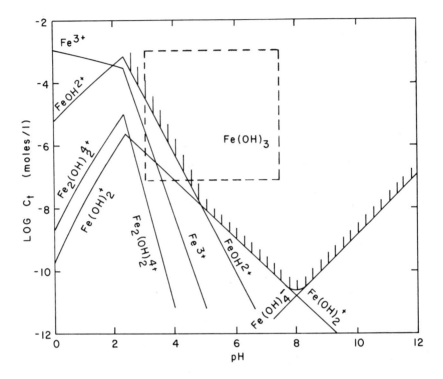

Figure 1.13. Log concentration-pH solubility diagram for ferric hydroxide.

in flotation activation, adsorption and water quality experiments is indicated by the enclosed area that spans both the soluble and insoluble metal hydroxy complex domains.

Adsorption, electrokinetic and coagulation experiments have all contributed to our knowledge of adsorption processes. Here, emphasis is given to adsorption data and their interpretation.

Adsorption of Aqueous Co(II)

To illustrate the adsorptive behavior of hydrolyzable metal ions at sediment/water interfaces, a typical divalent metal ion, cobalt(II), can be used as a case study to compare adsorption on some mineral components of sediments. For example, data is available for Co(II) adsorption on a variety of oxide minerals, e.g., a-quartz [93,94], manganese dioxides [95-98], geothite [94], rutile [93], amorphous iron oxide [99], gibbsite [94], and the clay mineral montmorillonite [100].

This selection of minerals is fairly representative of inorganic sediment components and provides a wide variety of inorganic surfaces. For example, the electrical surface charge properties range from the usually negatively charged silica to the positively charged alumina. The point of zero charge (PZC) of these minerals, as listed by Parks [24], is given in Table 1.4. A self-consistent set of hydrolysis data for aqueous Co(II) is given in Table 1.5.

Typical Co(II) adsorption data at solid/water interfaces are shown in Figures 1.14 and 1.15 as per cent adsorption from a known concentration added to the system and also as an adsorption density depending on the units in which the data were

Table 1.4

Surface Charge Properties of Some Sediment Components*

Mineral	Point of Zero Charge
a-quartz, SiO_2	2.5 (1.5-3.5)
SiO_2 gels	1.0-2.5
Mn(II) manganite	1.8
a-MnO_2	4.5
Geothite, a-FeOOH	6.7 (6.1-6.7)
$Fe_2O_3 \cdot XH_2O$	8.5
Gibbsite, a-$Al(OH)_3$ and all Al oxides and hydroxides	~ 9 (9.1-9.3)
Kaolin	~ 3.5 (< 2-4.6)
Montmorillonite	< 2.5

*After Parks [24].

Table 1.5

Self Consistent Set of Hydrolysis Data for Cobalt(II) at 25°C, I = 0*

No.	Equilibria	K	log K
1.	$Co(OH)_2(s) = Co^{2+} + 2OH^-$	K_{30}	-14.9
2.	$Co^{2+} + H_2O = Co(OH)^+ + H^+$	$*K_1$	-9.6
3.	$Co(OH)^+ + H_2O = Co(OH)_2^0 + H^+$	$*K_2$	-9.2
4.	$Co(OH)_2(s) = Co(OH)_2^0$	K_{S2}	-5.7
5.	$Co(OH)_2 + H_2O = Co(OH)_3^- + H^+$	$*K_3$	-12.7

*From Sillén and Martell [34].

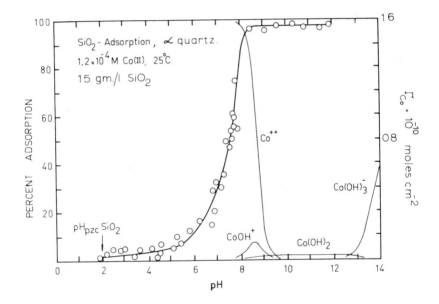

Figure 1.14. Experimental adsorption isotherm for Co(II). Adsorption at 1.2×10^{-4} M on silica at 25°C. The available surface area is 75 m²/l. Computed hydrolysis data for this concentration are also shown as the percentage of each aquo complex as a function of pH (after James and Healy [110]).

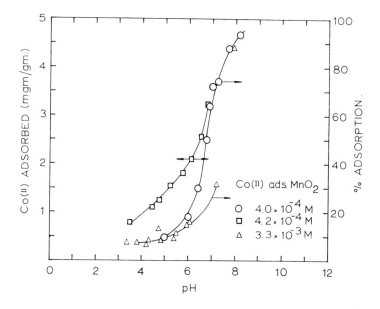

Figure 1.15. Experimental adsorption isotherms of Co(II) adsorption on MnO_2 as a function of pH. Data are plotted as weight per weight or percent adsorption. (After McKenzie [95], Tewari, et al. [96], and Loganathan [97].

reported. These data have been published during the last 20 years. In few cases were the thermodynamically significant adsorption density (moles/unit area) reported. Hence, it is unfortunate that we do not have an absolute measure of the efficiency of adsorption in terms of a dimensionless reaction constant or a free energy of adsorption.

If, however, the percentage of removal or adsorption from solution is compared, there is a remarkable similarity in the behavior of all the systems. That is, at low pH the percentage of adsorption (and the adsorption density) is low; however, when the pH is increased so that hydrolysis begins to become significant, the percentage adsorption increases from near 0 to 100%. Provided that the ratio of metal concentration to available surface area is great enough, reversal of the ζ potential is also observed near this pH region [93].

It appears that the enhanced removal of metal ion from solution may be due to the precipitation of solid hydrolysis products. However, the reaction is not this simple because most combinations of pM and pH shown here are well undersaturated with respect to solid hydroxide phases. Also, the concentration dependence of adsorption, namely, smaller percentage adsorption with increased added metal concentration, is opposite to the concentration dependence of precipitation and polymerization. That is, if polymerization and precipitation were directly responsible for the abrupt increase in fractional adsorption, the adsorption edge would move to lower pH values with increased metal concentration. The reason for this was outlined in the section on hydrolysis and is discussed more fully elsewhere [88].

There are some indications that hydrous iron and manganese oxides control the equilibrium concentration of heavy metals in natural aquatic systems [101]. The results shown here seem to indicate there may be nothing very special about the adsorption properties of these oxides (*e.g.*, exceptionally high adsorption energies) apart from their common abundance and very high surface areas. The results are also typical of the reaction of hydrolyzable metal ions in many aqueous colloidal dispersions.

Phenomenological Models for Interactions

In the analysis of the adsorption of metal ions on mineral and organic surfaces in natural water, it appears that three models may be applied to the description of the phenomena. The metal ions may be exchanged for surface protons [102] or the specific adsorption of certain monomeric or polymeric hydrolyzed species into the electrical double layer (EDL) may take place, modifying the surface properties of the mineral. Another model, proposed by James and Healy [88,93,110], explains that the preferential adsorption of soluble hydrolyzed metal ions by oxides is due to the high solvation energies of the unhydrolyzed ions opposing the attractive electrostatic and chemical adsorption interactions. At higher pH values and total metal concentrations, surface nucleation and precipitation of metal hydroxides leads to charge reversal of the mineral from negative to positive. This model describes the adsorption density or percentage adsorption of soluble species as a function of the metal ion and its properties, the solid phase and its properties and the pH, composition and ionic strength of the aqueous phase [110].

Ion Exchange Model

The ion exchange model is useful conceptually for interpreting the removal of aqueous metal ions by oxide surfaces. In aqueous dispersions the oxide surfaces become hydrated and in acid solutions the surface groups are chiefly hydroxylated, *i.e.*, *S*-OH with an excess of protonated sites, *S*-OH$_2^+$.

It has been proposed that $\equiv$S-O$^-$ can function as a ligand for metal ions M^{n+} which exchange for H^+ ions. For example, Duggar, *et al.* [102] proposed that the reaction of aqueous metal ions at the surface of silica gel is given by:

$$m(Si\text{-}OH) + M^{n+} \underset{}{\overset{K_{exch}}{\rightleftharpoons}} Si\text{-}O_m M^{(n-m)+} + mH^+ \qquad (28)$$

and that m is usually equal to n.

Measurement of changes in pH and pM allows calculation of stability constants for this reaction for several metals. The value of the free energy per *Si*-O-M bond (or log K_{exch}) was calculated by Dugger, *et al.* [102] for 20 metals and was shown to depend linearly on the $\overset{*}{p}K_1$ for the hydrolysis of the metal ion. Hence, the relation between uptake of metal ions and their hydrolytic reactions could be explained.

However, the proton exchange or ion exchange theory is not entirely satisfactory in this form because:

1. It lacks general application to other colloid systems. For example, the proton exchange model cannot explain the uptake of hydrolyzable metal ions on silver halides or polymer lattices because there are, presumably, no surface-bound protons available for exchange. Yet, metal ions show the same characteristic pH-dependent behavior in all of these systems.
2. It is a constant surface charge model; therefore, it can neither explain the observed changes in ζ potential nor the charge reversal when hydrolysis occurs. The only way in

which these observations could be explained is that $n > m$ in Equation 28 and that n/m is a function of pH.
3. Equation 28 may be written in a different but equivalent way that gives a different mechanistic interpretation, namely:

$$S] + M^{n+}_{aq} \rightleftharpoons S]M(OH)_m^{(n-M)+}(H_2O)_x + mH^+ \qquad (29)$$

In other words, adsorption of hydrated, hydrolyzed species is an alternative model with the same stoichiometry. It will be shown later that because the uptake of metals can be described by Equation 29 the ion exchange models and adsorption models are difficult to differentiate experimentally and to some extent both models can be applied.

Adsorption in the Double Layer

Many current studies have established that the adsorption of metal ions in the EDL is directly related to the hydrolytic behavior of the ion concerned. Some workers have interpreted their observations of abrupt changes in adsorption density or ζ potential as due to the adsorption of:

1. the first hydrolysis product, e.g., $PbOH^+$ on quartz [81], $ThOH^{3+}$ on AgI [103], and $LaOH^{2+}$ on AgI
2. specific ionic polynuclear hydrolysis products, e.g., $Al_8(OH)_{20}^{4+}$ on silver halide sols [105]
3. mononuclear nonelectrolyte complexes, e.g., $Hf(OH)_4^°$ on glass, PVC latex and AgI sols [106-108]
4. amorphous polymeric surface compound formation, e.g., $Al(OH)_3(s)$ on kaolinite [109].

While there is a wide difference of opinion on which of the hydrolyzed species is surface active, there is general agreement that the abrupt increase in adsorption, charge reversal and colloid coagulation is not due to the free aquo metal ion.

Hence, the theory of specific adsorption of certain hydrolysis products is successful, therefore, in predicting the approximate solution conditions necessary for high adsorption densities and charge reversal. However, it has rarely been quantified in terms of adsorption density as a function of concentration of ionic species, adsorption free energies and other system parameters.

At best, the predictions of the theory can only be described as qualitative. One of the problems with the specific adsorption models is that adsorption densities calculated using an electrostatic surface potential and a chemical adsorption potential in standard adsorption isotherms are usually overestimated.

General Model for Adsorption of Hydrolyzable Metal Ions

A recent model proposed by James and Healy [93,110] is an attempt to analyze the adsorption data for hydrolyzable metal complexes at solid/liquid interfaces in terms of our present knowledge of the structure of the EDL. This model is based on the concept that the solid/water interface is comprised of the solid, an adsorbed layer of interfacial water having some special properties, and bulk water. The manner in which these components govern the distribution of ionic species between the interface and the bulk solution can be evaluated and then expressed as an adsorption isotherm. For example, if the solid is an oxide, a surface charge is acquired by adsorption of H^+ or OH^- ions (or dissolution of M^{n+} or $O^=$ from the oxide lattice). The surface potential of the charged solid with respect to the bulk may be approximately written

$$\psi_o = \frac{RT}{F} (pH_{PZC} - pH) \quad \text{volt} \qquad (30)$$

Hence, we can consider how ionic species are distributed between the solution and the interface where the potential ψ is a function of ψ_o, position and ionic strength of the electrolyte solution.

A schematic summary of the model is shown in Figure 1.16; more detail concerning adsorption models is given by Grahame [111]. The simple Gouy and Chapman model (Figure 1.16A) distributes the ions between the solution and the interface according to the simple coulombic adsorption free energy, $\Delta G°_{coul} = ZF\psi$ joule mole^{-1}. Apart from the well-known limitations of this model for interfacial systems [111], it is most unsuitable for description of increased adsorption of hydrolyzed metal ions because $ZF\psi$ becomes smaller in magnitude, that is,

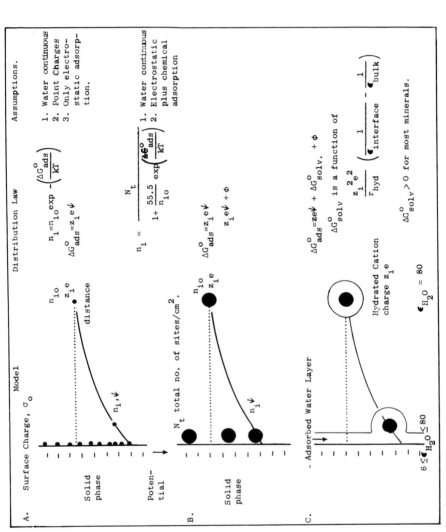

Figure 1.16. Schematics for phenomenological models for adsorption at the solid/solution interface.

less favorable to adsorption as Z decreases, while observed adsorption increases.

The refinement of EDL models due to Stern, shown in Figure 1.16B, has been particularly useful for describing the adsorption of ionic surfactants and anions because of the allowance for the possibility of "chemical" or noncoulombic interactions at the interface that enhance adsorption as well as the coulombic free energy of adsorption. Here, the total adsorption energy is given by $ZF\psi + \phi$, where ϕ is the specific adsorption free energy that is negative in sign. However, this type of distribution law is still not generally satisfactory for the description of hydrolyzable metal ions. It would still predict that the unhydrolyzed species adsorb strongly at the negatively charged oxide/water interfaces, even if ϕ was not operative for the unhydrolyzed metal ion.

To describe hydrolyzable metal ion adsorption we need an additional term in the free energy of adsorption of the Stern or Langmuir equation that opposes adsorption, *i.e.*, is positive in sign, of highly charged free aquo ions. It must also decrease in magnitude as the average charge z_i of the aqueous metal complexes decreases with hydrolysis. Such a term has been incorporated by James and Healy [110] (Figure 1.16C), who proposed that the change in solvation (or hydration) energy of the aqueous metal ions on adsorption, ΔG°_{solv}, contributes a significant, positive free energy to the total free energy of adsorption. That is,

$$\Delta G^\circ_{ads_i} = \Delta G^\circ_{coul_i} + \Delta G^\circ_{solv_i} + \Delta G^\circ_{chem_i} \tag{31}$$

where it can be shown [110,147] that for the adsorption of a hydrated cation, ΔG°_{solv}, is a function of

$$\left[\frac{z_i^2 e^2}{r_{hyd}}\right] \left[\frac{1}{\epsilon_{interface}} - \frac{1}{\epsilon_{bulk}}\right]$$

where z_i is the sign and charge of ion i, r_{hyd} is its hydrated radius and $\epsilon_{interface}$ and ϵ_{bulk} are the dielectric coefficients of the interfacial region and the bulk solution, respectively. For adsorption of ions onto all clay minerals and insulating oxides, i.e., low $\epsilon_{\text{interface}}$, then $[1/\epsilon_{\text{interface}} - 1/\epsilon_{\text{bulk}}]$ is positive and, hence, $\Delta G°_{\text{solv}}$ is positive.

By considering all soluble species as potential adsorbates, as for example the system illustrated in Figure 1.17, and using this

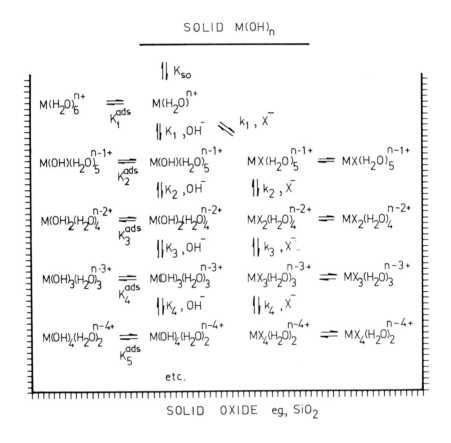

Figure 1.17. Matrix of competing equilibria involving aquo, hydroxide, and some other ligand (X) complexes, in solution, adsorbed at an interface, or involved in precipitate formation (after James and Healy [110]).

total free energy of adsorption, ΔG°_{solv} in a Langmuir-type equation, namely

$$\Theta_i = \frac{K_i M_i}{1 + \Sigma K_i M_i} \quad \text{and} \quad \Theta = \Sigma \Theta_i \quad (32)$$
$$\text{(total all species)}$$

where

$$K_i = \exp\left[-\frac{\Delta G^\circ_{ads_i}}{RT}\right] \quad (33)$$

and Θ_i is fractional coverage of surface by species i of equilibrium concentration M_i and adsorption energy, ΔG°_{ads}, James and Healy [110] were able to obtain quite good fit to experimental isotherms for Co(II), Fe(III), Cr(III) and Ca(II) on SiO_2 (shown in Figure 1.18) and Co(II) on TiO_2.

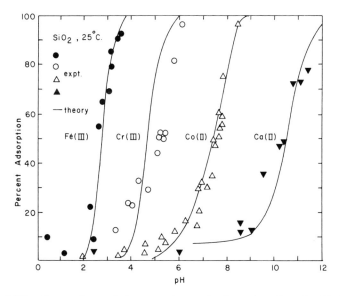

Figure 1.18. Comparison of the experimental adsorption data for Fe(III), Cr(III), Co(II), and Ca(II) on SiO_2 as a function of pH with the amount computed from the model. The added concentrations of metals are 1.2×10^{-4}, 2.0×10^{-4}, 1.2×10^{-4}, and 1.4×10^{-4} M, respectively (after James and Healy [110]).

54 Aqueous-Environmental Chemistry of Metals

In the absence of complexing ligands other than OH⁻, the total adsorption density, or the adsorption of individual species on a general inorganic sediment can be obtained from the mass balance equation:

$$[M]_{total} = [M^{n+}] \left[\sum_{\text{all species}} \beta_i [OH^-]^i + \frac{S(\Sigma K_i \beta_i [OH^-]^i)}{1 + [M^{n+}](\Sigma K_i \beta_i [OH^-]^i)} \right] \quad (34)$$

where $[M]_{total}$ is the total analytical concentration of the metal in aqueous solution, $[M^{n+}]$ and $[OH^-]$ are the concentrations of the free aquo complex and hydroxide ions, respectively. β_i is the stability constant of the hydroxy complexes, K_i is the adsorption coefficient of these species, and S is the total number of adsorption sites per unit volume. Solution of this equation for M^{n+} allows calculation of all the desired quantities, for example, equilibrium concentrations, adsorption densities and percent adsorption.

Effect of Complexing Ligands on Adsorption

An additional ligand in the system might affect adsorption in at least two ways. It might compete with OH⁻ in the coordination of the metal ion and repress the formation of hydroxo complexes, *i.e.*, it may alter the pH at which hydrolysis products adsorb. Or, it might form a second suite of adsorbing complexes. The model can account for this by including the other ligand complexes as solution species and potential adsorbates in the mass balance equation:

$$[M]_{total} = [M^{n+}] \left[\Sigma \beta_i [OH]^i + \Sigma \beta_j [L]^j + S \frac{\Sigma K_i \beta_i [OH]^i + \Sigma K_j \beta_j [L]^j}{1 + [M^{n+}][\Sigma K_i \beta_i [OH]^i + \Sigma K_j \beta_j [L]^j]} \right]$$

$$(35)$$

where β_i and β_j are the stability constants of hydroxo [OH⁻] and other ligand complexes [L⁻]; and K_i and K_j are the adsorption constants or coefficients calculated from the free energy of adsorption. S is the total number of adsorption sites per unit volume. This may be estimated from the cross sectional area of a hydrated metal ion and the total surface area dispersed per unit volume. The adsorbability of the ligand complexes can be estimated by variation of $\Delta G_{chem(ligand)}$ to give best fit with the experimental data. This influences the magnitude of adsorption coefficients, K_i and K_j, in Equation 35. Since both the hydroxo and chloro complexes of Hg(II) are relatively stable, the influence of pH and complexing ligand, Cl$^-$, on adsorption has been explored experimentally and an attempt to apply the above model to the adsorption reactions has been made.

Complex Equilibria of Mercury

Mercury(II) is well known for its ability to form stable complexes with inorganic ligands. Compared to most divalent metal ions Hg is strongly hydrolyzed. A set of hydrolysis data chosen from Sillén and Martell [34] is shown in Table 1.1. The corresponding log concentration-pH diagram for the solubility of HgO in water is presented in Figure 1.19. The maximum concentration of the neutral Hg(OH)$_2^\circ$ given by $K_{S2} = 10^{-3.7}$ M is more than an order of magnitude higher than the solubilities of most divalent heavy metal hydroxides. The solubility of HgO in the pH range that Hg(OH)$_2$ predominates is comparable with the solubility of a-quartz when $H_4SiO_4^\circ$ predominates.

Mercury(II) forms strong complexes with the halides in the order Cl$^-$ < Br$^-$ < I$^-$. The stability constants for the chloro complexes are also given in Table 1.1 [112,113]. The log concentration-pH diagrams for the chloro and hydroxo and hydroxo-chloro complexes are given in Figure 1.2A and 1.2B for the equilibrium chloride concentrations of 10^{-1} M and 10^{-3} M, respectively, and a total Hg(II) concentration of 10^{-7} M. The maximum concentration of HgCl$_2$ is given by $K_{S2} = 10^{-0.57}$ M [26].

From these figures it can be seen that over the concentration range [Cl$^-$] = 0 to 10^{-1} M and between pH 3 and 14, the predominant soluble mercury(II) species is either Hg(OH)$_2^\circ$ or HgCl$_2^\circ$. As the Cl$^-$ concentration is increased from 0 through 10^{-3} to 10^{-1}, Hg(OH)$_2^\circ$ becomes predominant at pH 3, 6.8, and 8.8.

56 *Aqueous-Environmental Chemistry of Metals*

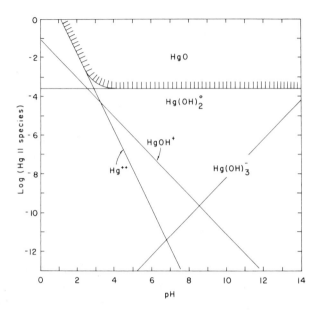

Figure 1.19. Log concentration-pH solubility diagram for HgO in aqueous solution at 25°C.

Adsorption of Hg(II) Complexes on Silica

The adsorption of 1.84×10^{-7} M Hg(II) from aqueous solutions of 0.1 ionic strength ($NaClO_4$, $NaClO_4$ + NaCl, and NaCl) onto silica dispersions is shown in Figure 1.20 and 1.21. These results correspond to the inclusion of zero, 10^{-3} M and 10^{-1} M Cl^- in the background electrolyte. It can be seen that for these conditions the percentage adsorption (or the adsorption density) increases abruptly in the pH ranges 2 to 3, 6 to 7, and 8 to 9. Below these pH values there is no detectable adsorption from solution. If these results are compared to the data presented in Figures 1.2 and 1.19, the increase in adsorption density correlates with the formation of $Hg(OH)_2^o$ as the predominant species.

There also appears to be a maximum in each isotherm that varies with the chloride concentration. The pH of the maximum increases with increasing chloride. In order to test whether this effect was due to increasing adsorption of Na^+_{aq} at the quartz/SiO_2

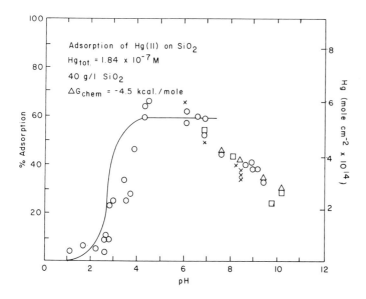

Figure 1.20. Experimental adsorption isotherms for Hg(II) from 1.84×10^{-7} M solutions by quartz as a function of pH and ionic strength. The ionic strength conditions are 10^{-1} M $NaClO_4$, 1×10^{-2} M $NaClO_4$, 2×10^{-3} M $NaClO_4$, 1×10^{-1} M $Mg(NO_3)_2$.

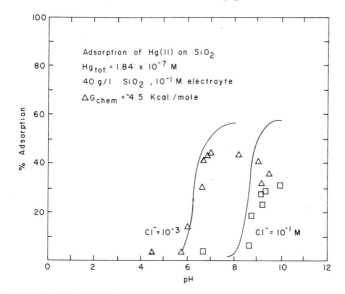

Figure 1.21. Experimental adsorption isotherm for Hg(II) from 1.84×10^{-7} M solutions by quartz as a function of pH and complexing ligand. The total ionic strength is 1×10^{-1} ($NaClO_4$, NaCl) where Cl^- is 1×10^{-3} M and 1×10^{-1} M.

interface [114,115], the effect of ionic strength [$NaClO_4$ and $Mg(ClO_4)_2$] of background electrolyte is shown in Figure 1.20. Within the experimental limits of pH imposed by the ionic strengths used, the form of the adsorption behavior of Hg(II) at the SiO_2/water interface is not affected by electrolyte concentration.

From the adsorption results, it appears that mercury(II) exhibits the general adsorption properties of hydrolyzable metals, e.g., the strong pH dependence of adsorption. The adsorption isotherms shown in Figures 1.20 and 1.21 show abrupt increases in adsorption for each isotherm. The pH region at which this occurs increases with chloride ion concentration. Qualitatively, the pH range of the abrupt increase corresponds to the pH range over which $Hg(OH)_2^\circ$ assumes dominance over $HgCl_2^\circ$. This behavior is summarized in Table 1.6. It can be inferred from these observations that Hg^{2+} and the Hg(II)-Cl$^-$ complexes are either not adsorbed at all or are only weakly adsorbed in comparison to the hydroxo complexes of Hg(II).

Quantitative interpretation of these observations is possible. Both hydroxo complex adsorption and ion exchange models apply to some extent as will be discussed in the next section. The parameters used in the James and Healy adsorption model are listed in Table 1.6. The line in Figure 1.20 results when the ΔG_{chem} term for the free metal and soluble metal hydroxo species in Equation 25 has the value -4.5 Kcal/mol. For the calculated isotherms in Figure 1.21, ΔG_{chem} for Hg_{aq}^{2+} and its hydrolysis products is also -4.5 Kcal/mol. However, ΔG_{chem} for the Hg(II)-Cl$^-$ complexes is 0 Kcal/mol.

The model calculations describe the adsorption behavior quite well in each case with only one general exception. This exception is the gradual decrease in adsorption once the pH is greater than the abrupt adsorption edge. There is a gradual decrease in adsorption density for pH values greater than 5. In order to test whether this decrease was due to the increased adsorption of Na^+ competing with $Hg(OH)_2^\circ$ species, adsorption experiments were carried out at lower ionic strengths (10^{-2} M and 2×10^{-3} M $NaClO_4$). Under these conditions, the adsorption density of Na^+ is much smaller than at 10^{-1} M $NaClO_4$ [114]. However, from Figure 1.20 it can be seen that there is no apparent dependence of adsorption of Hg(II) on ionic strength in the pH

Table 1.6

Adsorption of Hydrolysis Products of Hg(II) on SiO_2

Cl^- M	pH Range of Abrupt Adsorption Increase	pH at Which $Hg(OD)_2^\circ$ Becomes Predominant
0	2 – 3	3.0 $Hg^{2+} = Hg(OH)_2^\circ$
10^{-3}	6 – 7	6.8 $HgCl_2^\circ = Hg(OH)_2^\circ$
10^{-1}	8 – 9	8.8 $HgCl_2^\circ = Hg(OH)_2^\circ$

Model Parameters

$r_{Hg} = 1.1$ Å $\qquad r_{H_2O} = 2.76$ Å

Area per Hg(II) complex 50 Å$^2 \qquad$ 1:1 electrolyte

Ionic strength 10^{-1}

α-Quartz, SiO_2

Dielectric constant $SiO_2 = 4.3$

$pH_{PZC}(SiO_2) = 3.0$

Surface area: .5 m^2/gm and 40 gm/liter

ΔG_{chem}, $Hg(OH)_i^{2-i} = -4.5$ Kcal mole^{-1}

ΔG_{chem}, $HgCl_j^{2-j} = 0$ Kcal mole

---

range of these later experiments. Additional experiments were carried out in 10^{-1} M $Mg(NO_3)_2$ background electrolyte. Again, there is no significant difference in adsorption behavior. Thus, it appears that there is no exchange or competition between Na^+ or Mg^{2+} and the hydrolyzed Hg(II) species. This pH effect has also been observed for Hg(II) adsorption on other oxide/water interfaces, e.g., α-FeOOH, α-Al$_2$O$_3$ [116,117] and δ-MnO$_2$ [116]. Similar pH effects are also evident for Hg(II) adsorption on activated carbon [118], glass [119], and polyethylene [120].

The only ways that this observation could be described by the James and Healy model are: (1) the maximum number of adsorption sites, S, shows a gradual decrease with pH, (2) the chemical adsorption free energy $\Delta G_{chem,hydroxo\ species}$ contains terms that decrease in magnitude with pH, or (3) there may be other solution species not included in this analysis that adsorb less strongly than $Hg(OH)_2^{\circ}$.

Similar observations have been made for the adsorption of other dilute metals at oxide/water interfaces [121]. As yet, no solution is available; however, this problem will almost certainly confront environmental chemists dealing with very dilute heavy metal solutions, and further study of this aspect of adsorption is worthy of attention.

Relationship Between Adsorption and Ion Exchange

In the discussion of the Hg adsorption the data have been analyzed in terms of an adsorption model where the adsorption density, Γ_M, is proportional to the concentration of all soluble metal species in solution. However, the results may also be interpreted in terms of a modified ion exchange model in which only the free aquo metal ion exchanges for surface protons.

At least three models to account for the removal of aqueous metal ions from solution are conceivable. These include the ion exchange reaction:

$$m\ SiOH + M^{n+} \xrightleftharpoons{K_{exch}} SiO/_m\ M^{(n-m)+} + mH^+ \qquad (36)$$

the adsorption and hydrolysis at the surface reaction:

$$surface/ + mH_2O + M_{aq}^{n+} \rightleftharpoons surface/\left[M(OH)_{aq}^{(n-m)+}\right] + mH^+ \qquad (37)$$

and hydrolysis followed by an adsorption reaction:

$$M_{aq}^{n+} + mH_2O \xrightleftharpoons{^*\beta_m} M(OH)_m^{(n-m)+} + mH^+ \qquad (38)$$

and

$$\text{surface/} + M(OH)_m^{(n-m)+} \underset{}{\overset{K_{ads}}{\rightleftarrows}} \text{surface/}\left[M(OH)_m^{(n-m)+}\right] \quad (39)$$

The combination of reactions 38 and 39 results in reaction 37; thus, 38 plus 39 is thermodynamically undistinguishable from 37. Hence, the stability constant for both reaction 37 and reaction 38 plus 39 is $*\beta_m K_{ads}$. Also, because the stability constants for reaction 36, K_{exch}, and reaction 37, $*\beta_m K_{ads}$, have the same form, the exchange and adsorption models are indistinguishable unless there is some additional evidence, other than the metal concentration, pH or adsorption density, concerning the type of bonding at the surface. This may be verified by application of the more different models to the experimental data to check that either

$$K_{exch} = *\beta_m K_{ads} \quad (40)$$

or that the free energy changes for the reactions are comparable.

Using Equation 36 and following the method and assumptions of Dugger, et al. [102], the exchange constant K_{exch} was calculated for each experimental point from the equations

$$K_{exch} = \frac{1}{4}\left[\frac{1}{\Theta^2} - 1\right] \cdot \frac{a_H^2}{a_{Hg}} \quad (41)$$

where

$$a_{Hg} = [Hg_{aq}]\{1 + \Sigma*\beta_i/[H^+]^i + \Sigma\beta_j[Cl^-]^j + \Sigma\beta_{ij}[OH^-]^i[Cl^-]^j\}^{-1} \quad (42)$$

which reflects the effect of complexing ligands on metal ion activity and assuming

$$\Theta = 1 - 2\Gamma_{Hg}/\Gamma_{H_{max}} \quad (43)$$

i.e., Θ is the fraction of surface in the hydrogen form.

The calculations yield the value $\log_{10} K_{exch} = -3.2 \pm 0.6$. Using this value for K_{exch} and reversing the calculation procedure, the fractional removal of Hg(II) from aqueous solution by the silica was obtained. The shape and position of the curves from these calculations are very similar to those obtained in the adsorption model and shown in Figures 1.20 and 1.21. Inserting K_{exch} and $*\beta_2$ into Equation 40 yields the value of the adsorption constant for $Hg(OH)_2^°$, $K_{ads} = 10^{+2.8}$. Hence, $\Delta G_{ads} Hg(OH)_2^° = -1.38 \log_{10} K_{ads} = -3.9$ kcal/mol. This agrees quite well with the value of $\Delta G_{ads} = -4.5$ kcal/mol used in the adsorption model. The difference probably arises from the different stoichiometry of surface sites in the exchange and adsorption models.

Data for the solid/liquid distribution of metal ions obtained from ion exchange investigations may appear to be in conflict with data measured in direct adsorption investigations. It has been shown above that the mechanisms cannot be simply resolved. Some apparent differences in the experimental observations probably arise because (1) hydrolysis according to Equation 38 may have been neglected, and (2) there may be differences in the ranges of variables and which variables are measured. For example, on the one hand, ion exchange is inferred from pH changes associated with the uptake of metal ions added to the solid/liquid system. Since experiments are conducted at relatively low pH compared to the $p*K_1$ of the metal, the adsorption densities are low. On the other hand, in adsorption studies where the pH is systematically varied from low to high pH values resulting in abrupt increases in the metal ion uptake, the stoichiometry of the pH change required to induce the adsorption increase is often ignored.

In this discussion the attempt has been made to demonstrate the importance of the equilibrium solution species in determining the adsorption behavior of aqueous metal ions. It is evident from this and other studies, that the hydroxo complexes do have specific, noncoulombic interactions of a kind strongly favorable to adsorption, whereas other ligand complexes may not. These results are of general importance in the evaluation of the processes that occur when suspended oxides and clays bearing adsorbed heavy metals leave a fresh water environment for a marine environment. When this happens, adsorbed inorganic mercuric complexes may be liberated from oxide materials to be complexed by Cl⁻ and free to participate in other reactions and transport modes. When, as a result of further studies, we are able to learn more about the chemistry responsible for $\Delta G°_{chem}$ and ion exchange energies, we will be able to make some fairly reliable predictions of the adsorption of various metal ions at the solid/liquid interface.

It appears from the results at hand that the $\Delta G°_{chem}$ of hydroxy complexes of a given metal does not vary greatly from one solid oxide to another. Hence, in conclusion, it can probably be said that for the adsorption of hydrolyzed metal ions, the best substrate among the common inorganic components of sediments is probably the one with the highest total surface area.

TRANSPORT PROCESSES IN NATURAL AQUATIC SYSTEMS

The complex interactions and related processes controlling or influencing the behavior of trace metals in natural aquatic systems presents us with a challenging set of problems. If we are to ever understand and eventually model the behavior of trace metals in complex aquatic systems phenomenologically then we must gain insight into the diverse and dynamic mechanisms and processes influencing trace metals in natural systems. The necessary insight will be gained through a combination of laboratory experiments and field observations.

A description of the observed behavior of mercury in natural waters follows as a case study in the transport processes affecting trace metals. Other metals could be presented just as well, however, mercury has the advantage of being somewhat more complex chemically thus providing a somewhat more complete set of possible interactions.

Transport of Mercury

Mercury is fairly unique among heavy metals in that it is being continuously moved back and forth between all phases

of the environment, including air. It is well known that in
natural waters with organic as well as inorganic suspended
matter metal ions (free and complexed) are primarily found
associated with the hydrous oxides [117,122], clays [122] and
detrital and biological organic matter [78,123]. Mercury in
suspended sediments is reported to be 5 to 25 times the amount
in the filtered water [124-126]; however, neither the chemical
species of mercury nor the type of associated suspended material
are known. Apparently then, solute mercury species introduced
into streams are quickly transformed chemically or physically,
under the proper conditions, by sorption and surface complexation on inorganic and organic sorbates, and by sorption and
ingestion by viable biota.

Available observations indicate that suspended and bed
sediments of streams remove a major percentage of mercury
introduced into streams [127-130]. The rate of removal has
been found to be a function of the existing redox conditions,
the amount of suspended sediment, stream discharge, and the
mineralogical-chemical nature of the sediment. There is a considerable body of literature on the transport and deposition of
heavy metals such as lead, zinc, copper and cadmium in natural
water systems [131-135] and the literature on the movement of
mercury in aquatic environments is growing. It appears that
mercury behaves similarly to other metals such as zinc and cadmium in being strongly sorbed by soils and sediments.

Where streams have cut across mercury-bearing deposits,
both solute and particulate mercury are released directly into
the fluvial environment. In places, both thermal and nonthermal
springs and mine drainage contribute significantly to the mercury
content of streams. For example, a regular decrease in mercury
content down the Paglia River in Italy below a mercury anomaly
has been observed by Dall'Aglio [127]. Bayev [130] has noted
that the mercury content of stream sediments (for streams
intersecting dispersion aureoles) varied according to the type of
sediment. The decrease in mercury concentration of the water
moving away from mineralization zones varied with sediment
type: more rapid decrease for sandstone and clay sediments
compared to sediments containing predominantly limestone and
marl, presumably because of rapid sorption on clay particles.

Since mercury may enter the aquatic and terrestrial environment, and hence enter soils and sediments, in many different forms (*e.g.*, metallic Hg°, complexed or free Hg(II), alkyl-Hg or alkoxic-alkyl-Hg compounds), there appear to be a number of possible retention mechanisms that allow the metal to accumulate in sediments. Mercury must be fixed, that is, be desorbed very slowly, by soils and sediments. Otherwise, the high vapor pressure of free mercury and some of the organomercuric compounds would preclude the notable enrichment demonstrated by some soils situated over mercury deposits and the very considerable increase in mercury concentration in fluvial sediments immediately below industrial outfalls that are known to contain mercury wastes.

Mercury is known to form stable complexes with a number of different types of organic compounds found in natural waters, such as sulfur-containing proteins and humic materials. Sherbina [136,137] has demonstrated the important role of these complexes in the migration and precipitation of elements such as gold, uranium, and titanium in supergene zones. Many elements accumulate as simple cations in neutral surficial zones where they are not mobile because they are easily hydrolyzed and can precipitate as hydroxides. These elements, however, do migrate geochemically; the determining factor is apparently the formation of chelates with natural organics. Analogous phenomena with mercury can be expected.

Böstrom and Fisher [138] studied the distribution of mercury in East Pacific sediments and concluded that adsorption of mercury on colloidal matter was one alternative explanation for the distribution pattern they found. Krauskopf [122] studied the adsorption of 13 metals and found mercury most strongly adsorbed by marine plankton, hydrated ferric oxide, montmorillonite and peat moss. He concluded that surface sorption might be the controlling mechanism for transport and removal of mercury from the oceans. Klein and Goldberg [139] found the mercury content of sediments near an ocean outfall as much as fifty-fold higher than presumably uncontaminated sediments further away. Levels in sediments in a part of the Detroit River very close to a chlor-alkali plant were found as high as 86 ppm and decreased to 0.4 ppm and less near the river mouth at Lake Erie [140].

It is clear that interfacial phenomena play a major role in the distribution of mercury and other trace metals. Studies to date document the role of particulates in the transport process. None of the studies, however, have attempted to identify the form of the mercury in the sediments in any systematic way. However, the chemical form of mercury and the manner in which it is held in the sediments is pertinent to any attempt to estimate the probable reactions and/or movement of mercury through the aquatic environment and thus allow modeling of the basic processes.

Relative Residence Time of Mercury

It is appropriate to discuss the concept of a residence time (the time an atom remains in a system) which can be applied to a steady-state model of natural water systems [28]. For a lake or estuary, the following relationship holds

$$(d[Hg]_t/dt)_{inflow} = (d[Hg]_t/dt)_{sedimentation} + (d[Hg]_t/dt)_{outflow} \tag{44}$$

where $[Hg]_t$ is the total mercury concentration, whatever the form. For most water systems it is convenient to think in terms of a residence time, t_{rel}, relative to the residence time of water, t_{H_2O}:

$$t_{rel} = t_{Hg}/t_{H_2O} = \frac{[Hg]_t}{(d[Hg]_t/dt)_{inflow} \times t_{H_2O}} \tag{45}$$

The relative residence time, t_{rel}, is approximately 1 for conservative elements, which do not undergo any reactions (*e.g.*, adsorption, surface complexation, precipitation, etc.). An example is chloride. Reactive substances, however, have residence times different from those of water. Here t_{rel} can be either greater or smaller than unity. Elements that are readily incorporated into sediments and become unavailable to the solution will have t_{rel} less than 1 (*e.g.*, Al). On the other hand, mercury that is removed from the epilimnion of a lake as adsorbed or precipitated species may desorb or redissolve at the bottom under appropriate conditions. For example, reducing conditions in sediments may cause the partial release of sorbed or coprecipitated mercury on manganese and iron oxides due to the reductive dissolution of these solids. This would have the effect of increasing the abundance of soluble mercury by making more mercury available for forming organo-mercury complexes.

Conversely, it is possible that under suitable reducing conditions the mercury will be reduced to metallic mercury. Because of this cycle, t_{rel} for Hg will be larger than 1. This model includes as a part of the system that portion of the sediment that remains accessible to the overlying water. However, those atoms of mercury which get deeply buried and are not a dynamic part of the sediment-water interface would have essentially left the system.

Elements such as mercury which participate in biogeochemical cycles, have relative residence times that are usually much greater than unity. Through bio-uptake, algae may incorporate mercury into their biomass and eventually sink to the hypolimnion (or be consumed by primary predators and move up the food chain) where the organic matter is degraded and the metal would again accumulate in the lower layers of the lake. During the spring and fall turnover, the mercury may again become redistributed through the epilimnion and be available for bio-uptake. Hence, the mercury becomes trapped in the lake. Therefore, the concentration of mercury in dynamic natural water systems cannot be predicted merely from the inflow of the metal and the hydrographic conditions.

The extent of the accumulated mercury reservoir in a body of water may be more strongly affected by its relative residence time than by its supply to the body of water. The relative residence time, in turn, is influenced by the biota, by the mixing relationships, and by the exchange with the sediments. Furthermore, because many organisms have large bioconcentration factors for mercury, the potential for contamination of biota is not primarily a function of the soluble mercury concentration, but appears to depend rather on the rate of supply of the metal to the biosphere where it is assimilated. This rate is a composite function of the rates of (1) the biological turnover, (2) exchange with sediments, and (3) supply of mercury to the body of water.

Temporal and Spatial Distribution of Mercury

The interrelationships between physical processes (sedimentation, advection, convection), biological activity, and chemical interactions determines the temporal and spatial distribution of mercury. Two main mechanisms appear to be responsible for the progressive accumulation of mercury in bottom sediments: (1) adsorption or coprecipitation on inorganic and detrital organic particulates and (2) the preferential accumulation of mercury by some benthic organisms.

Possible seasonal variations of mercury remain unstudied to date. It is known that many elements have seasonal variations. For example, the seasonal variations in the concentrations of iron and phosphate in hypolimnetric waters usually occur together [141].

There are no quantitative data yet, but there may be considerable exchange of mercury between sediments and the adjoining aqueous phase in rivers, lakes, and estuaries. The distribution of mercury between sediments and the overlying water is of considerable importance when the effect of bioconcentration is considered. A dynamic equilibrium of mercury at a sediment-water interface may be represented by:

$$\text{Hg in aqueous phase (small fraction)} \rightleftarrows \text{Hg in sediment phase (large fraction)} \quad (46)$$

Questions of some interest are: to what extent is there a mercury reservoir in the sediment? How available is such a reservoir to the overlying waters? As noted, significant quantities of mercury have been found in some sediments.

Biological influence on mercury accumulated in sediments probably falls into three categories: (1) biological mediation of reactions changing the form of mercury, thus influencing transport by either releasing or tying it up, (2) physical effect of burrowing animals on the exchange of fluid between interstitial and overlying waters, and (3) accumulation of mercury in biotissue and export or import of mercury either in viable organisms or as organic detrital material. In any case, it is probable that aquatic biota will affect the vertical and horizontal movement of mercury in both the water column and the sediment.

In many respects, mercury behaves the same as zinc in aquatic environments, especially in terms of bioconcentration. Pomeroy, et al. [131] studied the flux of Zn^{65} through a salt-marsh ecosystem. They found that bacteria in the sediments competed for the zinc and reduced the amount taken up by the sediments. The Zn^{65} was adsorbed rapidly by sediments and bacteria and from there moved slowly into the biota over a period of days and months. They found that the metal entered the salt-marsh biota at the level of bacteria, filter-feeding organisms, and sediment-feeding organisms rather than at the level of primary producers (except via dinoflagellate blooms). They also observed that animal populations that consumed sediments, either as food or incidentally, removed zinc from the sediments. The ingestion of bacteria from the sediments by sediment-feeding animals is probably an important step in the flux from sediments to biota. Pomeroy, et al. [131] found the physical and biological transport processes to be of equal importance in the movement of zinc. A similar study of mercury in natural systems would be useful.

Many investigators have foreseen the dilemma posed by accumulation of mercury in sediments. If, in fact, the mercury cycle is to be broken, something must be done with the mercury reservoirs now in the sediments. Before any rational decision can be made it is necessary to have an estimate of the rate at which mercury will move from sediments back into the overlying waters. It is quite possible that cases exist where little or nothing needs to be done, while other situations may call for some remedial action. This type of determination awaits more detailed and quantitative insight into the processes and interactions occurring within the sediments and at the sediment-water interface.

SUMMARY

A functional description of the interrelated biogeochemical phenomena occurring in a natural water must include information on the processes and mechanisms responsible for the dynamic behavior of trace metals. Natural systems are very complex. Consequently, it is necessary to couple information from both the field and the laboratory in order to characterize the causal relationships effective in a natural water. It is the view here that phenomenological models are useful if they allow prediction, but even more important didactically, models provide a scientific metaphor that allows us to frame questions necessary to gain greater insight into natural phenomena.

It is clear that ultimately a process and mechanism orientation is necessary to the understanding of controls on trace metals in natural waters. The approach taken in this paper has been one of attempting to describe the behavior of trace metals in aquatic systems through phenomenological models. The state-of-the-art is such that we are still trying to ask the correct questions in many cases. Nevertheless, the availability of powerful analytical tools and computational methods should allow for great progress in the near future toward gaining a more quantitative insight into the dominant processes extant in natural aquatic systems.

REFERENCES

1. Reiche, P. *A Survey of Weathering Processes and Products.* (Albuquerque: University of New Mexico Publications in Geology Number 3, University of New Mexico Press, 1950).
2. Keller, W. D. *The Principles of Chemical Weathering.* (Columbia, Missouri: Lucas Brothers Publishers, 1957).
3. Jackson, T. A. and W. D. Keller. *Amer. J. Sci., 269,* 446 (1970).
4. Alloway, B. J. and B. E. Davies. *Geoderma, 5,* 197 (1971).
5. Kobayashi, J. *Proc. 5th Intern. Water Pollut. Research Conf., I-25/1* (1971).
6. Bertine, K. K. and E. D. Goldberg. *Science, 173,* 233 (1971).
7. Weiss, H. V., M. Koide and E. D. Goldberg. *Science, 174,* 692 (1971).
8. Bowen, H. J. M. *Trace Elements in Biochemistry.* (New York: Academic Press, 1966).
9. Weiss, H. V., M. Koide, and E. D. Goldberg. *Science, 172,* 261 (1971).
10. Joensuu, O. I. *Science, 172,* 1027 (1971).
11. Koksoy, M. and P. M. D. Bradshaw. *Colo. Sch. Mines Quart., 64,* 333 (1969).
12. Snyder, R. B., *et al.* Illinois Institute for Environmental Quality, Report HEQ 71-7 (1971), p. 23.
13. Wallace, R. A., *et al.* Oak Ridge National Laboratory, Oak Ridge, Tenn., ORNL NSF-EP 1 (1971).

14. United States Geological Survey. USGS Prof. Paper 713 (1970).
15. Jonasson, I. R. and R. W. Boyle. *Can. Min. Metal Bull., CIM* (January 1972).
16. Ettinger, M. B. U.S. Public Health Service Publ. 1440 (1966).
17. Mink, L. L., R. E. Williams, and A. T. Wallace. Idaho Bureau of Mines and Geology, Moscow, Idaho, Pamphlet 149 (1971).
18. Leckie, J. O. Unpublished data. Stanford University (1972).
19. Hunt, J. P. and H. Taube. *J. Chem. Phys., 18,* 757 (1950).
20. Hunt, J. P. and H. Taube. *J. Chem. Phys., 19,* 602 (1951).
21. Rutenberg, A. C. and H. Taube. *J. Chem. Phys., 20,* 825 (1952).
22. Robinson, R. A. and R. H. Stokes. *Electrolyte Solutions.* (London: Butterworths, 1965).
23. Morel, F. and J. J. Morgan. *Environ. Sci. Technol., 6,* 58 (1972).
24. Parks, G. A. *Che. Rev., 65,* 177 (1965).
25. Douglas, B. E. and D. H. McDaniel. *Concepts and Models of Inorganic Chemistry.* (Blaisdell Publishing Co., 1965), Chapter 6.
26. Butler, J. N. *Ionic Equilibria.* (Reading, Mass.: Addison-Wesley, 1964).
27. Sillén, L. G. in *Treatise on Analytical Chemistry,* I. M. Kolthoff and P. J. Elving, Eds. (New York: Interscience, 1959).
28. Stumm, W. and J. J. Morgan. *Aquatic Chemistry.* (New York: Wiley-Interscience, 1970).
29. Morgan, J. J. in *Equilibrium Concepts in Natural Water Systems.* (Washington, D.C.: American Chemical Society, Advances in Chemistry Series 67, 1967), p. 1.
30. Hem, J. D. in *Trace Inorganics in Water.* (Washington, D.C.: American Chemical Society, Advances in Chemistry Series 73, 1968), p. 98.
31. Sillén, L. G. in *Chemistry of the Coordinate Compounds.* (London: Pergamon Press, 1957), pp. 176-195.
32. Sillén, L. G. *Quart. Rev., 13,* 146 (1959).
33. Sillén, L. G. *Pure Appl. Chem., 17,* 55 (1968).
34. Sillén, L. G. and A. E. Martell. The Chemical Society, London, Spec. Publ. No. 17 (1964).
35. Sillén, L. G. and A. E. Martell. The Chemical Society, London, Spec. Publ. No. 25 (1971).
36. Latimer, W. M. *Oxidation Potentials,* 2nd ed. (New York: Prentice-Hall, 1952).
37. Wagman, D. D., W. H. Evans, V. B. Parker, I. Harlow, S. M. Bailey, and R. H. Schumm. National Bureau of Standards Technical Note 270-3 (1968).
38. Leckie, J. O. and W. Stumm. in *Advances in Water Quality Improvement-Physical and Chemical Processes,* E. Glouna and W. Eckenfelder, Eds. (Austin: University of Texas Press, 1970).
39. Hood, D. W., Ed. *Organic Matter in Natural Waters.* (University of Alaska: Institute of Marine Science Occasional Publication No. 1, 1970).

40. Faust, S. J. and J. V. Hunter. *Organic Compounds in Aquatic Environments.* (New York: Marcel Dekker, 1971).
41. Christensen, J. J., J. O. Hill, R. M. Izatt. *Science, 174,* 459 (1971).
42. Duursma, E. K. in *Organic Matter in Natural Waters,* D. W. Hood, Ed. (University of Alaska, Institute of Marine, Science, Occasional Publication No. 1, 1970), p. 387.
43. Duursma, E. K. in *Chemical Oceanography,* J. P. Riley and G. Skirrow, Eds. (New York: Academic Press, 1965).
44. Slowley, J. F. and D. W. Hood. *Geochim. Cosmochim. Acta, 35,* 121 (1971).
45. Alexander, J. E. and E. F. Corcoran. *Limnol. Oceanog., 12,* 236 (1967).
46. Shapiro, J. *J. Amer. Water Works Assoc., 56,* 1062 (1964).
47. Slowley, J. F., L. M. Jeffrey, and D. W. Hood. *Nature, 214,* 377 (1967).
48. Hodgson, J. F., L. W. Lindsay, and J. F. Trierweiler. *Soil Sci. Soc. Amer. Proc., 30,* 723 (1966).
49. Duce, R. A., J. G. Quinn, C. E. Olney, S. R. Piotrowicz, B. J. Ray, and T. L. Wade. *Science, 176,* 161 (1972).
50. Siegel, A. in *Organic Compounds in Aquatic Environments,* S. D. Faust and J. V. Hunter, Eds. (New York: Marcel Dekker, 1971), p. **265.**
51. Schnitzer, M. and H. Kodma. *Science, 153,* 70 (1966).
52. Schnitzer, M. in *Organic Compounds in Aquatic Environments,* S. D. Faust and J. V. Hunter, Eds. (New York: Marcel Dekker, 1971), p. 297.
53. Schnitzer, M. and S. I. M. Skinner. *Soil Sci., 102,* 361 (1966).
54. Malcolm, R. L., E. A. Jenne, and P. W. McKinley. in *Organic Matter in Natural Waters,* D. W. Hood, Ed. (University of Alaska, Institute of Marine Science, Occasional Publication No. 1, 1970), p. 479.
55. Stumm, W. *Proc. Third Intern. Conf. on Water Pollution Res.* (1966).
56. Morris, J. C. and W. Stumm. in *Equilibrium Concepts in Natural Water Systems.* (Washington, D.C.: American Chemical Society, Advances in Chemistry Series 67, 1967), p. 270.
57. Morgan, J. J. and W. Stumm. *Proc. Second Intern. Water Pollution Conf.* (1964), p. 103.
58. Berner, R. A. *Principles of Chemical Sedimentology.* (New York: McGraw-Hill, 1971).
59. Berner, R. A. *Amer. J. Sci., 267,* 19 (1969).
60. Berner, R. A. *J. Geol., 72,* 293 (1964).
61. Berner, R. A. *Amer. J. Sci., 265,* 773 (1967).
62. Gaudin, A. M., D. W. Fuerstenau and G. W. Mao. *Trans. A.I.M.E., 214,* 430 (1959).

63. Gaudin, A. M., D. W. Fuerstenau, and M. M. Turkanis. *Trans. A.I.M.E.*, 208, 65 (1957).
64. Fuerstenau, D. W. and P. H. Metzger. *Trans. A.I.M.E.*, 217, 119 (1960).
65. Schwarzenbach, G. and M. Widmer. *Helv. Chim. Acta*, 49, 111 (1966).
66. Freyberger, W. L. and P. L. deBruyn. *J. Phys. Chem.*, 61, 586 (1957).
67. Iwasaki, I. and P. L. deBruyn. *J. Phys. Chem.*, 62, 594 (1958).
68. Schwarzenbach, G. and M. Widmer. *Helv. Chim. Acta*, 46, 2613 (1963).
69. James, R. O. Unpublished data. Stanford University (1972).
70. MacNaughton, M. G. Unpublished data. Stanford University (1972).
71. Parks, G. A. in *Equilibrium Concepts in Natural Water Systems*. (Washington, D.C.: American Chemical Society, Advances in Chemistry Series 67, 1967), p. 121.
72. Healy, T. W., A. P. Herring and D. W. Fuerstenau. *J. Colloid. Sci.*, 21, 435 (1966).
73. Schindler, P. and H. R. Kamber. *Helv. Chim. Acta*, 53, 1781 (1968).
74. Stumm, W., C. P. Huang, and S. R. Jenkins. *Chim. Acta*, 42, 233 (1970).
75. Schindler, P. W. and H. Gamsjäger. *Kolloid-Z.*, 250, 759 (1972).
76. Tanford, T. in *Advances in Protein Chemistry*, Vol. 17, C. B. Anfinsen, Jr., M. L. Anson, K. Bailey, J. T. Edsall, Eds. (New York: Academic Press, 1962), p. 69.
77. Mortensen, J. L. *Soil Sci. Soc. Proc.*, 27, 179 (1963).
78. Saxby, J. D. *Rev. Pure Appl. Chem.*, 19, 131 (1969).
79. Neihof, R. and G. Loeb. *Dissolved Organic Matter in Seawater and the Electric Charge of Immersed Surfaces.* in press.
80. Garrett, W. D. in *Organic Matter in Natural Waters*, D. W. Hood, Ed. (University of Alaska, Institute of Marine Science Occasional Publication No. 1, 1970), p. 469.
81. Fuerstenau, M. C., D. A. Elgillani, and J. D. Miller. *Trans. A.I.M.E.*, 247, 11 (1970).
82. Parks, G. A. *Amer. Min.*, 57, 1163 (1972).
83. Smith, R. W. in *Nonequilibrium Systems in Natural Water Chemistry*. (Washington, D.C.: American Chemical Society, Advances in Chemistry Series 106, 1971), p. 250.
84. Bilinski, H. and S. Y. Tyree. in *Nonequilibrium Systems in Natural Water Chemistry*. (Washington, D.C.: American Chemical Society, Advances in Chemistry Series 106, 1971), p. 235.
85. Fuerstenau, D. W. *Pure Appl. Chem.*, 24, 135 (1970).
86. Mackenzie, J. M. W. *Trans. A.I.M.E.*, 235, 82 (1966).

87. Mackenzie, J. M. W. and R. T. O'Brien. *Trans. A.I.M.E.*, *244*, 168 (1969).
88. James, R. O. and T. W. Healty. *J. Colloid. Interfac. Sci.*, *40*, 42 (1972).
89. Clark, S. W. and S. R. B. Cooke. *Trans. A.I.M.E.*, *241*, 334 (1968).
90. Healy, T. W., R. O. James, and R. Cooper. in *Adsorption from Aqueous Solution.* (Washington, D.C.: American Chemical Society, Advances in Chemistry Series 79, 1968), p. 62.
91. Matijević, E. M. B. Abramson, R. H. Ottewill, K. F. Schultz, and M. Kerker. *J. Colloid. Interfac. Sci.*, *23*, 437 (1967).
92. O'Melia, C. R. and W. Stumm. *J. Colloid Interfac. Sci.*, *23*, 437 (1967).
93. James, R. O. and T. W. Healy. *J. Colloid Interfac. Sci.*, *40*, 53 (1972).
94. Tiller, K. G. Ph.D. Thesis, Cornell University (1965).
95. McKenzie, R. M. *Aust. J. Soil Res.*, *5*, 235 (1967).
96. Tewari, P. H., A. B. Campbell, and W. Lee. *Can. J. Chem.*, *50*, 1642 (1972).
97. Loganathan, P. Ph.D. Thesis, University of California, Davis (1971).
98. Murray, D. J., T. W. Healy, and D. W. Fuerstenau. in *Adsorption from Aqueous Solution.* (Washington, D.C.: American Chemical Society, Advances in Chemistry Series 79, 1968), p. 74.
99. Kurbatov, M. H., G. B. Wood, and J. K. Kurbatov. *J. Phys. Chem.*, *55*, 1170 (1951).
100. Hodgson, J. F. *Soil Sci. Soc. Amer. Proc.*, *24*, 165 (1960).
101. Jenne, E. A. in *Trace Inorganics in Water.* (Washington, D.C.: American Chemical Society, Advances in Chemistry Series 73, 1968), p. 337.
102. Dugger, D., L. J. Stanton, B. Irby, W. McConnell, W. Cummings, and R. Maatman. *J. Phys. Chem.*, *68*, 757 (1964).
103. Matijević, E., M. B. Abramson, R. H. Ottwill, K. F. Schultz, and M. Kerker. *J. Phys. Chem.*, *65*, 1724 (1961).
104. Abramson, M. B., M. J. Jaycock, and R. H. Ottwill. *J. Chem. Soc.*, *5034*, 5041 (1964).
105. Matijević, E., K. G. Mathai, R. H. Ottwill, and M. Kerker. *J. Phys. Chem.*, *65*, 826 (1961).
106. Stryker, L. J. and E. Matijević. in *Adsorption from Aqueous Solution.* (Washington, D.C.: American Chemical Society, Advances in Chemistry Series 79, 1968), p. 44.
107. Stryker, L. J. and E. Matijević. *J. Colloid Interfac. Sci.*, *31*, (1969).
108. Stryker, L. J. and E. Matijević. *Kolloid-Z.*, *233*, 912 (1969).

109. Hall, E. S. *Disc. Faraday Soc., 42,* 197 (1966).
110. James, R. O. and T. W. Healy. *J. Colloid Interfac. Sci., 40,* 65 (1972).
111. Grahame, D. C. *Chem. Rev., 41,* 441 (1947).
112. Partridge, J. A., R. M. Izatt, and J. J. Christensen. *J. Chem. Soc.,* 4231 (1965).
113. Ciavatta, L. and M. Grimaldi. *J. Inorg. Nucl. Chem., 30,* 197 (1968).
114. Allen, L. A. and E. Matijević. *J. Colloid Interfac. Sci., 31,* 287 (1969); *33,* 420 (1970); *35,* 66 (1971).
115. Li, H. C. and P. L. deBruyn. *Surface Sci., 5,* 203 (1966).
116. MacNaughton, M. G. Ph.D. Thesis, Stanford University (1973).
117. Shimomura, S., V. Nishihara, Y. Fukumoto, and Y. Tanase. *J. Hygiene Chem., 15,* 84 (1969).
118. Maillet, S. Unpublished data. Stanford University (1973).
119. Benes, P. *Coll. Czech. Chem. Comm., 35,* 1349 (1969).
120. Benes, P. *Coll. Czech. Chem. Comm., 34,* 1375 (1966).
121. Benes, P., and A. Garba. *Radiochim. Acta, 5,* 99 (1966).
122. Krauskopf, K. P. *Geochim. Cosmochim. Acta, 9,* 1 (1956).
123. Pillai, T. N. V., M. V. M. Desai, E. Mathew, S. Ganapathy, and A. K. Ganguly. *Current Sci., 40,* 75 (1971).
124. Hinkle, M. E. and R. E. Learned. USGS Professional Paper 650-D, D251 (1969).
125. Bothner, M. H. and R. Carpenter. *Proc. Internat. Atomic Energy Agency, IAEA/SM-158/5* (1972).
126. Cranston, R. E. and D. E. Buckley. *Environ. Sci. Technol., 6,* 274 (1972).
127. Dall'Aglio, M. in *Origin and Distribution of Elements.* L. H. Ahrens, Ed. (New York: Pergamon Press, 1968), p. 1065.
128. Heide, F., *et al. Naturwissenshaften, 44,* 441 (1957).
129. Stock, A. and F. Cucuel. *Naturwissenhaften, 22,* 390 (1934).
130. Bayev, Y. C. *Dokl. Acad. Nank. SSSR, 181,* 211 (1968).
131. Pomeroy, L. R., *et al. Proc. Internat. Atomic Energy Agency* (Vienna), 177 (1966).
132. Horowitz, A. *Marine Geo., 9,* 241 (1970).
133. Delfino, J. J., *et al. Environ. Sci. Technol., 3,* 1189 (1969).
134. Turekian, K. K. and M. R. Scott. *Environ. Sci. Technol., 1,* 940 (1967).
135. Presley, R. J., R. R. Brooks and I. R. Kaplan. *Science, 158,* 906 (1967).
136. Sherbina, V. V. *Geochimiya, 5,* 54 (1956).
137. Sherbina, V. V. *Geochimiya, 11,* 340 (1962).
138. Bostrom, K. and D. E. Fisher. *Geochim. Cosmochim. Acta, 33,* 743 (1969).

139. Klein, D. H. and E. D. Goldberg. *Environ. Sci. Technol.*, *4*, 765 (1970).
140. "Mercury Stirs More Pollution Concern," *Chem. Engr. News* (June 22, 1970), p. 36.
141. Hutchinson, G. E. *A Treatise on Limnology*, Vol. 1 (New York: John Wiley, 1957).
142. Bell, L. C., A. M. Posner, and J. P. Quirk. *Nature, 239,* 515 (1972).
143. Bell, L. C., A. M. Posner, and J. P. Quirk. *J. Colloid Interfac. Sci., 42,* 250 (1973).
144. Somasundaran, P. *J. Colloid Interfac. Sci., 27,* 659 (1968).
145. Somasundaran, P. and G. E. Agar. *Trans. A.I.M.E., 254,* 348 (1972).
146. James, R. O. and G. A. Parks. (to be published).
147. Levine, S. *Disc. Faraday Soc., 52,* (1971).

CHAPTER 2

SOURCES AND DISTRIBUTION OF TRACE METALS IN AQUATIC ENVIRONMENTS

Sherman L. Williams, Donald B. Aulenbach
and Nicholas L. Clesceri
    Hudson-Lake George Limnological Research Center, Inc.
    and Rensselaer Polytechnic Institute
    Troy, New York

Sources of Metals in Rivers and Lakes	78
Rock and Soil	78
Precipitation and Atmospheric Fallout	86
Pollution - Contamination	90
Forms and Reactions of Metals in Aquatic Systems	103
Distribution in Typical Natural Waters	108
Oligotrophic-Mesotropic Lakes	109
Mesotrophic-Eutrophic Lakes	114
Streams and Rivers	118
Summary	123

The first series of transition elements in the periodic table is made up of scandium, titanium, vanadium, chromium, manganese, iron, cobalt, nickel and copper, followed by zinc, a related metal. All but the first two of these metals play a biochemical role in the life processes of some if not all aquatic plants and animals; therefore their presence in trace amounts in the aquatic environment is essential. Even the first two elements, although having no recognizable essential function, are apparently incorporated in aquatic organisms in a degree similar to the other elements. Although most of these elements are essential to aquatic organisms in trace quantities, at high concentrations these metals become toxic. This chapter will discuss the introduction of these transition metals and zinc into the aquatic environment, and the various processes that appear to control their concentrations and distributions in lakes and rivers. The concentrations of some of these metals in oligotrophic, eutrophic and polluted lakes and rivers in and adjacent to the eastern Adirondack Mountain region of New York state are presented along with data from the literature to illustrate their occurrence in typical natural waters.

SOURCES OF METALS IN RIVERS AND LAKES

Transition metals enter rivers and lakes from a variety of sources. The rocks and soils directly exposed to surface waters are usually the largest natural source. Dead and decomposing vegetation and animal matter also contribute small amounts of metals to adjacent waters. Wet and dry fallout of atmospheric particulate matter derived from natural sources, such as the dust from the weathering of rock and soil as well as from man's activities, which include the combustion of fossil fuels and the processing of metals, can introduce relatively large quantities of trace metals to rivers and lakes. Other activities of man, including the discharge of various treated and untreated liquid wastes to the water body or drainage basin or the contamination of the drainage basin through various construction, mining, lumbering or similar activities, can introduce large quantities of trace metals to local streams or lakes. The major sources will be discussed individually since their relative importance will vary widely from location to location.

Rock and Soil

The rock and soil of the drainage basin is the most important natural source of metals to the water bodies present in the basin. The amount of transition metals and zinc present varies greatly with the rock type and mineral content. Table 2.1 shows the average concentrations found in the major rock types.

Different plutonic rocks formed from the same parent magma usually have different amounts of trace metals as well as major

Table 2.1

Average Concentration of Metals in Various Types of Rock and Deep Ocean Sediments*
(values in mg/kg)

	Ultramatic	Basaltic	Plutonic Granitic Plagiclase	Plutonic Granitic Orthoclase	Syenite
Scandium	15	30	14	7	3
Titanium	300	1.38×10^4	3.4×10^3	1.2×10^3	3.5×10^3
Vanadium	40	250	88	44	30
Chromium	1.6×10^3	170	22	4.1	2
Manganese	1.62×10^3	1.5×10^3	540	370	850
Iron	9.4×10^4	8.65×10^4	2.64×10^4	1.42×10^4	3.67×10^4
Cobalt	150	48	7	1.0	1
Nickel	2.0×10^3	130	15	4.5	4
Copper	10	87	30	10	5
Zinc	50	105	60	39	132

	Sedimentary Rock Shale	Sedimentary Rock Sandstone	Sedimentary Rock Carbonate	Deep Ocean Sediments Carbonate	Deep Ocean Sediments Clay
Scandium	13	1	1	2	19
Titanium	4.6×10^3	1.5×10^3	4.0×10^2	770	4.6×10^3
Vanadium	132	20	20	20	20
Chromium	90	35	11	11	90
Manganese	850	100	1.1×10^3	1.0×10^3	6.7×10^3
Iron	4.72×10^4	9.8×10^3	3.8×10^3	9.0×10^3	6.5×10^4
Cobalt	19	0.3	0.1	7	74
Nickel	68	2.0	20	30	225
Copper	45	1	4	30	250
Zinc	95	16	20	35	165

*from Turekian and Wedepohl [1].

constituents depending upon the thermodynamics of the magma crystallization process. The fate of a trace metal will depend upon its concentration in the parent magma and the nature of the structural lattice that forms. Table 2.2 illustrates the variation in concentration that will occur as the crystallization process progresses and different rocks are formed. The changes in concentration of the various trace metals follow a generally reproducible pattern predictable on the basis of their charge, ionic radius and electronegativity. These factors result either in certain trace metals replacing others or the more prevalent iron and titanium in crystal lattices, or in their capture or incorporation without replacement in other lattices.

In sedimentary rocks different processes were at work which resulted in concentrations and concentration ratios different from those found through crystallization processes. Clastic sediments, for example, which are formed by the transport and deposit of rock particles resulting from weathering, are generally more highly enriched in Ti, V, Cr, Fe, Co and Ni because these elements occur in the residual mineral grains remaining after physical and chemical destruction of the other rock constituents. Biogenic sediments formed by living organisms generally show selective enrichment of most transition elements. Enrichments vary greatly with the environment in which the sediment was deposited as well as the biogenic materials constituting the sediment. Table 2.3 shows some typical average concentrations of transition metals in the various marine sediments. Table 2.4 shows representative concentrations of metals in oil, asphalt and coal and their ash. The selective enrichment of metals in these natural fuel materials has had a profound effect on their return to the terrestrial and aquatic environment, particularly in the case of the relatively less abundant elements such as vanadium, nickel, copper and zinc.

Soils and clay minerals contain significant quantities of these metals. Iron is a major component of illitic and montmorillonitic clays (4.12 and 2.82% Fe_2O_3, respectively), and makes up 0.19% of kaolinitic clays [6]. Titanium is present in smaller amounts in all three clays (0.62% TiO_2, 0.76% TiO_2 and 0.08% TiO_2 in illitic, kaolinitic and montmorillonitic clay, respectively). Soils with large organic fractions are capable of

Table 2.2

The Content of Minor Elements in Differentiates of the Skaergaard Intrusion Compared to that in the Initial Magma* (values in mg/kg)

	Initial Magma	Gabbro Picrite and Eucrite (earliest)	Olivine Gabbro	Olivine-free Gabbro	Ferrogabbros	Hedenbergite Granophyre	Granophyre (latest)
Rb	-	-	-	-	-	30	200
Ba	40	25	25	45	50	450	1700
Sr	300	200	700	450	700	500	300
La	-	-	-	-	-	25	150
Y	-	-	-	-	-	175	200
Zr	40	40	35	25	20	500	700
Sc	15	7	20	15	10	-	-
Cu	130	70	80	175	400	500	20
Co	50	80	55	40	40	10	4
Ni	200	600	135	40	-	5	5
Li	3	3	2	3	3	25	12
V	150	170	225	400	15	-	12
Cr	300	700	175	-	-	-	3
Ga	15	12	23	15	20	40	30
Proportion of rocks as percentage		65	14	10	7.50	2.50	0.50

*from Wager [2] and Wager, et al. [3].

Table 2.3

Average Concentrations of Metals in Marine Sediments*
(values in mg/kg)

	TiO_2	V	Cr	MnO_2	Fe_2O_3	Co	Ni	Cu
3 Siliceous pelagic sediments from Atlantic and Pacific	6500	460	97	6100	64200	200	330	370
2 Pelagic siliceous sediments from Pacific	5900	-	-	6700	64400	-	-	-
10 Pelagic calcareous sediments from all major oceans	3800	300	51	4000	38900	91	232	338
15 Calcareous pelagic sediments from Pacific	1000	-	-	14900	43600	-	-	-
12 Argillaceous sediments from all major oceans	8400	390	55	5900	82300	100	300	400
9 Pelagic red clays from Pacific	8000	-	-	7100	78600	-	-	-
35 Pacific pelagic sediments	22000	450	93	19700	92900	160	320	740
18 Atlantic deep-sea sediments	8500	45	86	6300	82000	38	140	130
12 Near-shore sands, Gulf of Paria	4500	79	31	1700	46900	77000	16	71000
12 Clays, Gulf of Paria	7200	-	-	-	-	-	-	-
North Atlantic calcareous core	4000	13	36	3800	42200	28	138	-
8 Red clays from Pacific	12500	48	51	14900	-	89	164	-
20 Pacific pelagic red clays	-	145	82	-	-	98	260	-
Pacific pelagic core	-	30	-	-	-	-	-	261
Atlantic pelagic core	-	51	-	-	-	-	-	178
2 Near-shore muds ($<2\mu$ fraction)	-	230	260	-	-	0	29	440
6 Near-shore greenish muds, Gulf of Paria	-	146	93	3200	72600	12	31	17
29 Near-shore sediments	-	-	82	-	-	11	47	-
Oceanic pelagic average	5900	193	69	5000	61800	100	236	345
Oceanic pelagic average	7100	-	-	8900	65400	-	-	-

*adapted from Riley and Skirrow [4].

Table 2.4
Metals Content of Oil, Asphalt, Coal and Their Ash*
(values in mg/kg)

	Oil	Ash	Asphalt	Ash	Pelitic Sediments (coal)	Coal Ash
Sc	ND	ND	ND	ND	6.5	400
Ti	Trace	8,000	ND	ND	4,300	20,000
V	260	460,000	23,400	530,000	120	11,000
Cr	3	450	6	330	550	1,200
Mn	330	33,000	0.4	ND	620	22,000
Fe	850	530,000	420	250,000	-	-
Co	1	10,000	0.14	70	8	2,000
Nu	240	158,000	3,000	200,000	24	16,000
Cu	19	130,000	1,660	108,000	192	4,000
Zn	0.6	18,000	100	1,000	500	10,000

*from Krejci-Graf [5].

binding significant quantities of transition metals and zinc through organic-metal complexing. The strength of these bonds for different metals will vary with the pH. In addition, hydrous oxides of iron, manganese and titanium in the soils and clays will adsorb metals selectively.

Krieger [7] found that Cu, Mn and Zn are concentrated in the A_0 and A_1 horizons of podzolic soils by recycling by plants. He also noted:

1. that, on the average, chernozems and black prairie soils contain more copper than podzols and podzolic types, and that better drained soils contain more copper than do poorly drained ones
2. that black prairie soils and chernozems contain less manganese than do podzols and podzolic soils, and that it is more evenly distributed throughout profiles of the former
3. that black prairie soils and chernozems contain slightly more zinc than podzolic types, and that it is more evenly distributed throughout the former two soils.

Zhukova and Solodnikova [8] found that in soils exposed to frequent leaching and oozing regimes such as in irrigated regions, the upper layers contained highly active forms of Cu, average Co and low Zn. Mn was high in clayey soils, and in loamy soils was low in the spring and summer and high in late summer and fall. In these soils, Cu, Mn and Co decrease with depth while Zn increases. Jenne [9] concluded that the hydrous oxides of Mn and Fe furnish the principal controls for the fixation of Mn, Fe, Co, Ni, Cu and Zn in both soils and fresh water sediments.

The concentrations of metals in the nonmineralized zone of British soils in the Tamar Valley area are shown in Table 2.5.

Table 2.5

Average Metal Concentrations in Tamar Soils*

Metal	Concentration, mg/kg	
	Available	Total
Mn	101.6	-
Fe	42.6	-
Zn	4.31	64.39
Cu	0.50	22.09
Co	1.04	22.25
Ni	1.82	36.60

*from Davies [10].

In the table "available metals" are those leachable with dilute acetic acid, while "total" represents that extracted from the soil with hot concentrated nitric acid followed by hot aqua regia. These soils overlay Upper Devonian and Lower Carboniferous sedimentary rocks. By comparison, a pasture soil near an old mining area in the same vicinity contained 43.0 ppm available Zn, 16.9 ppm available Cu, 258 ppm total Zn, and 163 ppm total Cu, thus showing how the surrounding rocks can affect the metal content of adjoining soils.

Leaf and forest vegetation can also contribute significantly to soils as well as to adjoining water bodies. This contribution can be magnified when metal fallout from an air pollution source occurs. Little and Martin [11] report zinc levels up to 7000 ppm dry weight of plant material on the sheltered side of oak leaves near a lead and zinc smelter in Britain, with 6200 ppm on the exposed side. The acetic acid leachable (available) zinc in the pasture soils near the smelter ranged between 87 and 338 ppm in oven-dried soil, while the soil in adjacent woodlands ranged from 710 to 1100 ppm Zn. Even in relatively uncontaminated areas the concentration of metals in leaves and litter is appreciable. Nilsson [12] reported the ranges of concentrations of several trace metals in living spruce needles in the forest litter and in the underlying soil in a forest in southern Sweden, as shown in Table 2.6. The concentrations of some trace metals in various leaves from a New Hampshire forest were measured by Gosz, *et al.* [13] as shown in Table 2.7.

Table 2.6

Ranges of Concentrations of Trace Metals in Portions of Forest in Southern Sweden*
(values in mg/kg in dry matter)

Metal	Living Spruce Needles	Litter	Underlying Soil
Cr	0.2-1.5	2.3-6.7	23-81
Cu	1.4-3.5	5.1-13.5	37-48
Fe	29-111	574-2850	19,000-24,000
Mn	71-1090	131-318	290-323
Ni	0.7-3.1	2.7-5.6	18-28
Zn	27-188	60-126	28-41

*from Nilsson [12].

Table 2.7

Concentrations of Trace Metals in Various Leaves
from a New Hampshire Forest*
(values in mg/kg in dry matter)

Metal	Yellow Birch Leaves	Sugar Maple Leaves	Beech Leaves
Cu	9.3	6.7	7.0
Fe	109	83	164
Mn	4,538	2,759	2,299
Zn	509	31	37

*from Gosz, et al. [13].

Precipitation and Atmospheric Fallout

The second most important source of trace transition metals to water bodies on a global basis is the introduction through precipitation or dry fallout on the water surface or in the drainage basin. The air that moves over the earth's surface carries a large quantity of particulate matter or aerosols—about 10 million tons by one estimate [14]. This particular material is constantly being removed by gravitational settling and impingement on surfaces as well as by rain and snow almost as fast as it is being added. The material added to the atmosphere comes from natural sources as well as through man's activities and contains appreciable amounts of trace metals. Among the natural sources are the dust from volcanic eruption, the dust produced by the erosion and weathering of rocks and soil, the solids produced from the evaporation of entrained sea spray or other water droplets, and the smoke from forest fires and micrometeorite dust.

Man's activities, which account for about every one out of five particles or one out of every five tons in the air [15], add an even greater proportion of transition metals mainly because of the combustion of fossil fuels, which are enriched in trace metals, and the manufacture of steel and other metals. For example, it is estimated that over 25 million metric tons of Fe_2O_3 were discharged into the air from iron and steel manufacture in 1968 throughout the world, of which about 6 million tons were discharged in the United States [15]. A million metric tons of manganese oxide were also discharged as part of the steel manufacturing. In the case of vanadium, it is estimated by Zoller, et al. [16] that on an annual worldwide basis 27,000 metric tons enter the atmosphere from soil and rock dust, 10,000 metric tons from volcanic debris, 12 metric tons from sea salts, and 20,000 metric tons from the annual combustion of 2 billion metric tons of petroleum.

The particles containing the trace metals are returned to the earth's surface and water bodies mainly in four ways:

1. Gravitational settling returns most particles. This not only involves the settling of relatively large fly ash and soil particles but also the settling of smaller particles that collide and coagulate. Particles that are less than 0.1 micron in diameter move randomly in air, collide often with other particles, and thus grow rapidly by coagulation. Particles in this size range would soon vanish from the air were they not replenished constantly by condensation processes. Particles in the next larger size range, 0.1 to 1.0 micron, are also removed from the air primarily by coagulation. They grow more slowly than the smaller particles because they are somewhat less numerous, move less rapidly in air, and collide less often with other particles. At diameters larger than 1 micron, particles begin to develop appreciable settling velocities. Above 10 microns, they begin to settle relatively rapidly, although such particles can be kept airborne by turbulence for extended periods.
2. Particles brought near the earth's surface may be removed by impacting on the surface of obstacles such as vegetation, rocks and buildings.
3. Precipitation in the form of rain and snow returns particles by gathering large numbers of small particles in the cloud-forming process (in cloud scavenging), which are carried to the ground in the precipitation.
4. Precipitation can remove or wash out particles that are below the clouds during the precipitation event.

Deposited particulate matter settling out due to gravity has been measured in dust fall containers since the early 1900's, and as would be expected showed considerable local and temporal variation. Typical dry fallout in rural areas near large cities and industrial areas range between 10 and 100 tons per square mile per year, while in the thickly populated industrial areas, 500 to 2,000 tons per square mile per year are typical [14]. On occasion, the concentration of particles in the air over thickly populated industrial areas (experiencing the 500-2,000 ton fallout) has exceeded 1,000 $\mu g/m^3$ of air; however, in most urban areas the concentration averages about 105 $\mu g/m^3$. Suburban areas generally average about 37 $\mu g/m^3$ and remote continental areas 7 $\mu g/m^3$ [17,18]. In these suburban areas, dry fallout rates of 125 micrograms/m^2/day generally occur [19].

Precipitation will also remove particulate matter, the amount varying greatly with the amount and kind of precipitation. In suburban or rural regions most of the particles brought down with precipitation are collected in the cloud-forming process at the higher altitudes. Particles less than 2 microns in size are not effectively "washed out" by precipitation below the clouds. In rural areas with ample precipitation, the amount of material brought to the earth's surface by precipitation is about the same as that returned by dry fallout [20,21].

The concentration of transition metals and zinc will vary greatly in both wet and dry fallout, depending upon the source of the particles present in the area. Table 2.8 gives some typical metal concentrations in the air. The amount of dry fallout of

Table 2.8

Typical Concentrations of Various Metals in Air Particles in the United States and Adjoining Oceans* ($\mu g/m^3$ of air)

	Urban	Rural-Suburban	Oceanic
Total particles	105	37	11
Scandium	5×10^{-4}	1.1×10^{-4}	4.1×10^{-5}
Titanium	0.04	0.03	0.03
Vanadium	0.05	0.005	2.0×10^{-4}
Chromium	0.02	0.01	4.2×10^{-4}
Manganese	0.7	0.05	0.02
Iron	2.5	0.2	0.23
Cobalt	0.005	0.001	8.7×10^{-5}
Nickel	0.10	0.05	-
Copper	0.2	0.1	0.05
Zinc	0.7	0.05	0.0025

*from References 16, 17, 22-29.

these metals will be directly proportional to their concentrations in the air particles. For example, in a rural-suburban area where the nickel averages 0.1 $\mu g/m^3$ of air, a dry fallout of about 0.3 micrograms of nickel per square meter per day would be expected.

The removal of airborne particles and metals by precipitation will vary with the amount and type of precipitation. The concentration of metals in the precipitation will generally be greater during the first hour or so and may decrease with time. This is most noticeable if the precipitation is washing out particles below the clouds, particularly in contaminated atmospheres where there is a large particle concentration in the air close to the earth's surface. Table 2.9 lists the concentrations of these metals that have been measured in precipitation falling on undeveloped areas in the northwest and northeast (Quilayute, Wash., and Lake George, N.Y., respectively). The data bear out the

Table 2.9

Typical Concentrations of Various Metals in Rainwater ($\mu g/l$)

	Northwest[a] (rural)	Northeast[b] (rural)	Continental Av.[c]
Scandium	< 0.002	ND	ND
Titanium	ND	ND	ND
Vanadium	ND	ND	ND
Chromium	< 2	ND	ND
Manganese	2.6	3.42	12
Iron	32	34.89	NR
Cobalt	0.04	ND	ND
Nickel	ND	ND	43
Copper	3.1	8.16	21
Zinc	19.5	30.32	107

[a] from Rancitelli, et al. [26]
[b] from Williams, et al. [30] and unpublished data
[c] from Lazrus, et al. [31].

observations of Lazrus, et al. [31] who sampled the precipitation for Fe, Mn, Cu, Zn and Pb at 32 stations distributed throughout the United States including Albany, N.Y., and Medford, Oregon. They found lower concentrations of the metals in precipitation in the northwest than in the northeast. Of the first series of transition metals, only the concentrations of scandium, titanium and possibly cobalt and iron reflect concentrations in the atmosphere and in precipitation on a regional basis that could be completely attributed to natural sources of introduction [26].

Rahn [32] attributes the relatively high zinc and copper concentrations in the air over remote areas to man's activities. However, Zoller, et al. [33] could not ascribe the high Zn and Cu enrichment of Antarctica air particles to any natural or anthropogenic process. Vanadium, when present in detectable concentrations, usually results from combustion processes. The contribution of wet and dry fallout to the aquatic environment is expected to continue to increase as amounts of atmospheric particles increase. Significant increases in the concentrations of these particles have been taking place during the past 20 years over most of the United States, Europe and the North Atlantic.

Pollution-Contamination

The third principal source of trace metal introduction to the aquatic environment is man's pollution and contamination of the water bodies by direct discharge of various treated or untreated agricultural, municipal, residential or industrial effluents to the water body or through contamination or disturbance of the drainage basin with subsequent contaminated runoff to the water body. The contamination of the water body or drainage basin through the additional wet and dry fallout of metals, attributable to man's activities, has just been discussed.

Agriculture

Trace metals play an important part in the nutritional requirement for food and food products. Photosynthesis requires manganese, iron, chloride, zinc, and vanadium [34]. Also, for nitrogen fixation, iron, boron, molybdenum, and cobalt are required, and other functions may also require manganese, boron, cobalt, copper and silicon. The exact ratio of these metals in the plant material varies for each individual species. Some typical concentrations of trace elements found in plants are shown in Table 2.10. Furthermore, the uptake may be greater than the minimum requirement for that metal in the specific plant when excess amounts of the heavy metals are available. As noted previously, excessive concentrations of these trace metals may also result in inhibition or death of the plant grown on such soils. Concentrations which have been shown to be toxic to plants are also presented in the table.

Table 2.10

Total Concentrations of Trace Elements Typically Found in Plants and Soils

Element	Conc. in Plants ($\mu g/g$)		Conc. in Soils ($\mu g/g$)	
	Normal	Toxic[c]	Common	Range
Cr	0.2 -1.0	-	100	5-3000
Co	0.05-0.5	-	8	1-40
Cu	4-15	>20	20	2-100
Mn	15-100	-	850	100-4000
Ni	1	>50	40	10-1000
V	0.1 -10	>10	100	20-500
Zn	15-200	>200	50	10-300

[a]from Allaway [35].

[b]Toxicities listed do not apply to certain accumulator plant species.

Farm animals consume either the plant itself or its seeds. Here again trace amounts of these metals may be essential to the growth of the animals consuming the plant material. Furthermore, the concentration of the trace metals within the animal or within any particular organ of the animal may also vary. Excretion of the unmetabolized or waste metals may result in their accumulation and/or concentration in the waste disposal area. In the case of the plant material, this may be in the silage, whereas for animals it would be in the manure.

The normal source for these trace metals for plants is the soil in which they are grown. Typical concentrations are also shown in Table 2.10. The distribution of these metals in the rock and soils has been discussed earlier. In general, the concentrations available are very small, and in some cases may not be sufficient to provide the needed requirements of the plants. The solubility and consequently the plant uptake of the trace metals in the soils varies considerably with pH and the redox potential within the soil or the root systems [36]. The associated anion and the particle size also have a great influence on growth and metal uptake by plants. Precipitation may enhance solution, and precipitation and runoff may encourage erosion of the soil and rock to increase the availability of the trace metals.

Another source of trace metals is the water applied for irrigation. Little consideration has been given to the desirability of trace metals in irrigation water. However, maximum concentrations for various soils and application rates have been specified as shown in Table 2.11. For comparison, the specified limits for

Table 2.11

Surface and Irrigation Water Quality Criteria for Trace Elements (mg/l)

		Irrigation Water		
		Continuous Use		Short-Term Use
Element	Surface Water FWPCA[a]	FWPCA Any Soil	NAS[b] Coarse-Textured Soil	FWPCA[c] Fine-Textured Soil
Co	-	0.2	0.05	10.0
Cr	0.05	5.0	0.1	20.0
Cu	1.0	0.2	0.2	5.0
Mn	0.05	2.0	0.2	20.0
Ni	-	0.5	0.2	2.0
V	-	10.0	0.1	10.0
Zn	5.0	5.0	2.0	10.0

[a] U.S. Dept. of Interior, Federal Water Pollution Control Administration [37]. Surface water criteria are virtually the same as drinking water standards (U.S. Dept. of Health, Education, and Welfare [38]).

[b] National Academy of Sciences [39]. Recommended maximum concentrations of trace elements in irrigation waters used for sensitive crops on soils with low capacities to retain these elements in unavailable forms.

[c] For short-term use only on fine-textured soils.

trace metals in surface waters are also shown in this table. Not all of the irrigation water applied to a field remains because much runs off in order to maintain the salt balance within that field. The excess runoff may be similar to the quality of the applied water, and frequently has a higher concentration of some substances due to evaporation of the water and leaching of substances already present in the soil.

A manmade source of trace metals may be the fertilizers applied to the soil. Whereas the concentrations may vary in specific fertilizers, they have been shown to contain chromium, copper, iron, manganese, nickel, and zinc. The levels are seldom measured, however, and therefore little is known as to the total amount of trace metals contributed by fertilizers. In general, alkaline soils tend to stabilize trace metals. Thus, if lime is added along with fertilizer there is a greater chance of retention of the trace metals, possibly making them unavailable to the plant system. Phosphates also stabilize the metals, preventing them from becoming available to plants.

In general, the soils become a sink for the trace metals in that they are stabilized through oxidation reactions and adsorption. Obviously there is a limit to the chemical and adsorption reactions. Usually, water that percolates through soils is depleted of its heavy metals in a relatively short time. Thus, the trace metals have less chance of reaching a nearby stream. On the other hand, surface runoff may carry trace metals to a nearby stream.

It may be seen that the specific amounts of trace metals that reach receiving streams from agricultural runoff are extremely variable and difficult to evaluate. Very few stream surveys involving heavy metal determinations have been made; even fewer have been conducted comparing the specific effects of heavy metals in runoff from forested land, crop land, grazing areas and other agriculture-related industries. Therefore, specific values for the contribution of trace metals from the agriculture industry to streams is difficult to provide.

Domestic Sewage

Food containing the nutrients and trace metals taken up from the soil is consumed by humans and other animals. Not all the nutrients and trace metals are retained in the body tissues, the excess being passed off in the excrement, through perspiration, and in breathing. The first two reach the aquatic system directly where water carriage of wastes is employed. Significant problems can also arise where animals are concentrated, such as feed lots and chicken farms.

The actual amounts of trace metals in domestic sewage may vary according to water usage, quantity and types of food eaten, time of year, economic status, and the prevalence of garbage grinders. Studies in New York City revealed concentrations of metals in sewage as shown in Table 2.12.

Table 2.12

Trace Metals in New York City Sewage Treatment Plants*

Metal	Avg. Conc. in Treat. Plant Inf. mg/l	Avg. Conc. in Surface Runoff mg/l	Avg. Conc. in Treat. Plant Eff. mg/l	Treat. Plant Removal %	Avg. Conc. in Sludge mg/kg
Cu	0.27	0.46	0.15	45	60
Cr	0.16	0.16	0.08	48	38
Ni	0.11	0.15	0.10	17	8.1
Zn	0.41	1.6	0.26	36	85

*from Klein, et al. [40].

The data for similar studies at Grand Rapids, Mich., Richmond, Ind., and Rockford, Ill. are shown in Table 2.13. The values reported for Richmond were similar to those for New York City, whereas higher concentrations of all the metals measured were found at Grand Rapids and Rockford.

Recently, much attention has been given to the possibility of applying sewage and sewage sludge to the land. This has resulted in numerous studies of the trace metals concentrations in these sources. The studies by Kirkham [42], based on the average concentrations of metals in the sludges of other cities across the United States and Europe, indicate higher concentrations of metals than found in the New York City study. This is most

Table 2.13

Trace Metals in Sewage Treatment Plants of Three U.S. Cities*

City	Metal	Composite period, hr.	Sewage Total Metal		Primary Effluent Total		Primary Effluent Soluble		Final Effluent Total		Final Effluent Soluble	
			Avg.	Range	Avg.	Range	Avg.	Range	Avg.	Range	Avg.	Range
A	Cr	24	3.6	0.7-5.6	3.2	0.6-6.3	-	-	2.5	1.0-3.3	-	-
		8	3.8	0.6-5.1	3.5	0.6-5.3	2.8	0.3-4.0	2.6	1.0-3.8	1.7	0.2-3.1
	Cu	24	1.4	0.7-2.4	1.5	0.6-2.8	-	-	1.6	0.4-2.9	-	-
		8	1.6	0.3-3.7	1.4	0.4-2.3	1.4	0.5-2.7	1.6	0.3-3.2	1.3	0.2-2.6
	Ni	24	2.0	1.3-3.4	1.8	0.9-2.9	-	-	1.8	1.0-2.5	-	-
		8	2.1	1.2-3.5	1.9	1.0-2.4	1.7	0.8-2.2	1.8	1.0-2.2	1.6	0.8-2.1
	Zn	24	1.5	0.6-2.5	1.0	0.4-1.5	-	-	0.8	0.6-1.2	-	-
		8	1.5	0.4-2.2	1.0	0.4-1.6	0.2	0.1-0.4	0.7	0.6-0.9	0.3	0.2-0.6
B	Cr	24	0.8	0.2-2.1	0.8	0.3-1.8	-	-	0.2	0.01-0.5	-	-
		8	0.3	0.2-1.2	0.7	0.4-1.0	0.3	0.01-1.2	0.1	0.01-0.5	0.04	0.01-0.1
	Cu	24	0.2	0.1-0.4	0.3	0.2-0.6	-	-	0.07	0.04-0.2	-	-
		8	0.2	0.1-0.5	0.3	0.2-0.3	0.1	0.2-0.3	0.05	0.03-0.1	0.04	0.01-0.1
	Ni	24	0.03	0.01-0.1	0.03	0.01-0.05	-	-	0.02	0.01-0.03	-	-
		8	0.03	0.01-0.1	0.1	0.02-0.2	0.04	0.01-0.1	0.02	0.01-0.04	0.02	0.01-0.1
	Zn	24	0.3	0.1-0.5	0.4	0.3-0.9	-	-	0.1	0.1-0.2	-	-
		8	0.3	0.2-0.5	0.3	0.3-0.5	0.1	0.04-0.1	0.1	0.1-0.2	0.1	0.1-0.2
C	Cr	24	1.8	0.5-2.9	1.5	0.5-2.2	-	-	1.2	0.6-1.5	-	-
		8	2.7	0.4-3.9	1.8	0.3-3.0	1.1	0.1-2.5	1.2	0.2-1.8	0.6	0.05-1.5
	Cu	24	1.4	0.6-3.3	1.3	0.5-2.6	-	-	1.0	0.5-3.6	-	-
		8	1.7	0.4-6.8	1.7	0.2-7.5	1.5	0.2-5.5	1.6	0.4-7.3	1.2	0.2-5.5
	Ni	24	0.9	0.2-1.9	0.9	0.3-1.2	-	-	0.9	0.5-1.4	-	-
		8	1.0	0.08-2.2	1.0	0.3-1.6	0.9	0.2-1.4	1.0	0.5-1.9	1.0	0.6-1.5
	Zn	24	2.7	1.2-3.4	2.0	0.9-3.6	-	-	1.3	0.8-1.7	-	-
		8	3.7	1.0-10	2.1	0.8-3.4	0.5	0.2-1.2	1.4	1.0-1.8	0.2	0.1-0.5

A. Metals in process effluents: Grand Rapids, Michigan, September 1963. Concentrations for 14-day period, mg/l.
B. Metals in process effluents: Richmond, Indiana, August 1963. Concentrations for 14-day period, mg/l.
C. Metals in process effluents: Rockford, Illinois, October-November 1963. Concentrations for 13-day period, mg/l.
* from Public Health Service [41].

likely due to greater industrialization in those other cities. Balkeslee [43] showed (Table 2.14) that the range of concentrations of trace metals in sewage sludge varies with the treatment process. Argo and Culp [44] reported on a similar study. The results show that appreciable amounts of heavy metals may be contributed by domestic wastes, even more by combined domestic and industrial wastes, that activated sludge treatment removes less than 50% of these metals, and that these metals are concentrated in the sludge.

Table 2.14

Range of Trace Elements Found in Sewage Sludge in Relation to Treatment Process* ($\mu g/g$)

Element	Undigested Liquid Sludge	Secondary Digestor Sludge	Vacuum Filter Cake
Cr	66-7800	22-9600	28-10,600
Cu	200-1740	260-10,400	84-2600
Ni	44-740	14-1440	12-2800
Zn	900-8400	1120-16,400	480-9400

*from Blakeslee [43].

Runoff From Highway and Street Surfaces

In addition to the metals occurring in sewage, runoff from streets and highways can introduce significant quantities of metals into receiving waters. The amount of material and metal composition available for runoff will depend upon the street cleaning practices of the city or town and the activities on the street or the lands adjoining the street. Average concentrations of trace metals in surface runoff in New York City were shown as part of Table 2.12. Similar measurements made of the particulates washed from paved surfaces during artificial rainfall and expressed both in concentrations as well as amount per length of curb are summarized in Table 2.15. Table 2.16 compares metal-loading from road surface runoff to normal sanitary sewage for a typical

Table 2.15

Metal Concentrations in Particulates in Runoff from Paved Surfaces*

Metal	City Street mg/kg	City Street lb/curb mi	Rural Road mg/kg	Rural Road lb/curb mi	Highway mg/kg	Highway lb/curb mi	Airport mg/kg
Cr	209	0.23	215	0.34	185	1.20	125
Mn	440	0.47	860	1.35	370	2.39	310
Fe	24,000	24.4	23,000	36	21,000	136	21,000
Ni	34	0.04	105	0.16	105	0.68	85
Cu	120	0.13	39	0.06	40	0.26	18
Zn	400	0.41	70	0.11	190	1.24	75

*from Pitt and Amy [45]

city of 100,000 people during a one-hour rainfall of 0.1 inches of rain. The use of road salt and sand on streets and highways during the winter months in northern regions will also introduce additional metals in the runoff during periods of melting.

Table 2.16

Comparison of Metal Loadings in Road Runoff with Sanitary Sewage*

Metal	Road Runoff lbs/hr	Sanitary Sewage lbs/hr	Ratio Runoff/Sewage
Cr	80	12	6.7
Mn	160	9.7	15
Fe	7900	54	150
Ni	10	0.042	240
Cu	36	0.17	210
Zn	140	0.84	170

*from Pitt and Amy [45]

Mining and Mine Drainage

Another wastewater problem is the discharge of acid from active or abandoned mines, particularly coal mines [46]. Where coal is mined, the associated mineral, pyrite, containing iron and sulfur, is exposed to water and air. This pyrite breaks down to form sulfuric acid and acid-producing compounds of iron. Through pumping or by natural drainage the acid wastes flow into nearby streams. Unable to neutralize these acid wastes completely, many streams become unfit for use by man and can no longer support fish life. Both the Allegheny and the Monongahela Rivers are examples of acid streams.

The Allegheny River begins in north central Pennsylvania, flows into New York, reenters Pennsylvania, and meanders southward to Pittsburgh. In addition to the acid mine drainage, oilfield brines and mill wastes also are discharged to the headwaters of the Allegheny.

The headwaters of the Monongahela River are in northern West Virginia, from whence the river flow north to Pittsburgh. Here the Monongahela and Allegheny converge to form the Ohio River. It has been estimated that the Monongahela empties the equivalent of 180,000 metric tons of sulfuric acid each year into the Ohio. However, by the time the Ohio has passed the mouth of the Muskingum River, 170 miles (282 km) downstream, the acid load has been neutralized completely. Sulfuric acid, pH, iron, manganese, and sometimes aluminum are the parameters traditionally used for assessing the concentration of the mine drainage as well as its effect on a receiving stream. However, zinc, manganese, copper, nickel and cobalt are also frequently present [46].

Streams receiving acid mine drainage directly are referred to as primary streams. Two examples would be the Kiskiminitas River and the Youghiogheny River. Trace element concentrations in these two streams are given in Table 2.17.

The Kiskiminitas and the Youghiogheny flow into the Allegheny and the Monongahela Rivers, respectively. The latter are secondary receiving streams. Table 2.18 lists trace element concentrations found in the Allegheny and the Monongahela Rivers at Pittsburgh, Pennsylvania, before the two rivers converge

Table 2.17

Trace Element Concentrations in Two Rivers
Receiving Acid Mine Drainage (mg/l)*

	Kiskiminitas River at Apollo, Pa.							
Period	Zn	Fe	Mn	Cu	Ni	Co	Cr	V
4/29-5/28/65	430	>1,500	>1,500	100	55	23	<5	<20
5/29-6/28/65	265	>1,000	> 500	23	55	22	<3	<10
6/29-7/28/65	380	>1,000	>1,000	17	115	48	<5	<20
7/29-8/28/65	170	>1,000	>1,000	13	95	40	<5	<10
8/29-9/28/65	465	>2,500	>1,250	15	138	60	<6	<25
	Youghiogheny River at West Newton, Pa.							
2/28-3/28/65	70	>1,000	245	10	<10	<10	<5	<20
3/29-4/28/65	25	>1,000	175	8	<10	<10	<5	<20
4/29-5/28/65	50	>1,000	290	13	15	10	<5	<20
5/29-6/28/65	25	>1,000	215	3	<5	<5	5	<10
6/29-7/28/65	20	>2,000	280	3	<10	<10	<5	<20
7/29-8/28/65	40	>1,000	190	3	<5	<5	<3	<10
8/29-9/28/65	20	>1,000	225	3	<10	<10	<5	<20

*from Kopp and Kroner [46]

to form the Ohio. Zinc, iron, and manganese are detected with regular frequency. Although observed in both streams, zinc, manganese, and copper appear at much higher levels in the Monongahela. Cobalt and nickel also are detected frequently. The relatively high levels of zinc, iron and manganese observed at Pittsburgh apparently do not remain in solution for any length of time, as indicated by analyses of four samples taken at Toronto, Ohio, some miles downstream. Those elements associated with acid mine drainage occur in solution at much lower levels than at Pittsburgh, indicating that precipitation or absorption occurs.

Table 2.18

Trace Element Concentrations Observed at Pittsburgh, Pa.*

Element	Monongahela River (19 samples)			Allegheny River (20 samples)		
	Frequency of Detection(%)	Range (ppb)	Average (ppb)	Frequency of Detection(%)	Range (ppb)	Average (ppb)
Zinc	100	54-787	271	80	7-114	43
Iron	89	7-183	53	65	4-57	23
Manganese	100	12-1,288	665	85	1-1,000	284
Copper	95	3-157	47	30	2-5	4
Chromium	16	3-36	15	35	1-8	4
Nickel	89	10-86	32	40	8-57	30
Cobalt	47	7-34	14	20	3-12	7
Vanadium	5	-	43	10	2-11	7

*from Kopp and Kroner [46]

Land Clearing

Land clearing is primarily related to the lumbering industry and the construction industry; however, the clearing of land for agricultural purposes must also be considered in this category. In general the effect here is a secondary one. Whereas wastes from the lumbering industry may contribute heavy metals to the aquatic environment as described under the next section on industrial wastes, the prime cause for concern is the increase in erosion caused by the clearing of the land without proper reseeding of the formerly forested areas. In a similar manner, cleared land, particularly during the construction period, can be subject to extreme erosion. This erosion increases the potential for carrying of the heavy metals to the aquatic environment. No specific measurements could be found of the concentrations of heavy metals that could be contributed to the aquatic environment because of land clearing. Obviously, a great variation can occur due to the concentration of the trace metals in the land being cleared.

Industrial Discharges

Representative concentrations of the various metals in the effluents from industries are difficult to provide, since they will vary with the type of industry and with specific operations within that industry. Depending upon the processing operation, the same product may result in different amounts of trace metals being contributed to the aquatic environment. In addition, the amounts of water used will vary the concentration of the trace metals in the wastewaters. Frequently, where water conservation is practiced in an industry, the concentrations of the heavy metals are actually greater in the effluent. On the other hand, processes have been developed in which there is minimum of contact of water with the materials containing the metals. In these cases, the contribution of the metals to the aquatic environment is lessened. Furthermore, various treatment processes have been devised to treat the industrial wastes to remove the polluting material in general and the heavy metals in particular cases.

Very few measurements have been made for the specific effect of industrial discharges into a stream from the standpoint of the trace metals involved. However, some data were accumulated by Kopp and Kroner [46]. Table 2.19 lists summaries of trace element data on the Maumee, the Cuyahoga, and the St. Mary's Rivers in Ohio. The Maumee River travels through an industrial complex from which wastes from metal-working, petro-chemicals, and extensive agricultural operations are added before it reaches Toledo, Ohio, and flows into Lake Erie. The Cuyahoga, a similar example of a stream receiving much industrial waste, travels from Akron to Cleveland, Ohio, past a complex of automotive, paper, and meatpacking industries. The St. Mary's River, which connects Lake Superior and Lake Huron, receives little industrial waste and can be considered a clean stream. Trace elements were observed more frequently in the St. Mary's River but the concentrations were much lower.

Trace elements were measured in the Delaware River at Martins Creek, Pa., Trenton, N.J., and Philadelphia, Pa. All three stations were combined to show the results indicated in Table 2.20. Measurable concentrations of zinc and copper were observed in essentially all of the samples. Iron and manganese

Table 2.19

Trace Element Concentrations in Rivers Receiving Industrial Discharges*

Maumee River at Toledo, Ohio (16 samples)						
	Zn	Fe	Mn	Cu	Cr	Ni
No. of positive values	13	12	10	13	5	6
Frequency of detection (%)	81	75	63	81	31	38
Minimum value (ppb)	10	9	1.6	5	6	9
Maximum value (ppb)	96	56	16	39	15	58
Mean of positive (ppb)	30	24	6.4	10	11	25

Cobalt - all negative, <17 ppb; Vanadium - all negative, <34 ppb
Beryllium - one positive, 0.19 ppb; Cadmium - one positive, 9 ppb

Cuyahoga River at Cleveland, Ohio (16 samples)						
	Zn	Fe	Mn	Cu	Cr	Ni
No. of positive values	16	11	14	5	3	11
Frequency of detection (%)	100	69	88	31	19	69
Minimum value (ppb)	54	6	4.5	6	8	15
Maximum value (ppb)	1,183	312	900	14	15	120
Mean of positive (ppb)	423	59	285	9	11	70

Silver - all negative, <4.5 ppb; Vanadium - all negative, <60 ppb
Cobalt - one positive, 20 ppb; Beryllium - two positive, 0.16 ppb

St. Mary's River at Sault Sainte Marie, Michigan (15 samples)						
	Zn	Fe	Mn	Cu	Cr	Ni
No. of positive values	15	12	9	15	3	3
Frequency of detection (%)	100	80	60	100	20	20
Minimum value (ppb)	2	1	0.3	2	1	2
Maximum value (ppb)	406	168	4.0	28	7	24
Mean of positive (ppb)	46	24	1.8	6	3	11

Silver - all negative, <4 ppb; Vanadium - all negative, <7 ppb
Beryllium - all negative, <0.05 ppb; Cadmium - all negative, <10 ppb
Cobalt - one positive, 2 ppb

*from Kopp and Kroner [46]

Table 2.20

Trace Elements in the Delaware River (57 samples at 3 stations)*

Element	Frequency (%)	Minimum (ppb)	Maximum (ppb)	Average (ppb)
Zinc	99	3	178	36
Iron	84	1	195	21
Manganese	67	0.3	6.3	1.7
Copper	100	2	47	13
Chromium	23	2	29	7
Nickel	42	1	49	8
Cobalt	2			All < 9
Vanadium	4			All <14

*from Kopp and Kroner [46]

also occurred with regularity. In general, the Delaware River may be considered a rather highly industrialized river, and the trace metals represent, for the most part, the contributions from industrial discharges.

FORMS AND REACTIONS OF METALS IN AQUATIC SYSTEMS

Metals in natural waters may be suspended, colloidal or soluble. In general, suspended particles are considered to be those greater than 100 mμ in size, soluble are those less than 1 mμ with colloidal particles falling in the intermediate range. The suspended and colloidal particles may consist of individual or mixed metals in the form of their hydroxides, oxides, silicates, sulfides or as other compounds, or they may consist of clay, silica, or organic matter to which metals are bound by adsorption, ion exchange or complexation. The soluble metals may be ions, simple or complex, or un-ionized organometallic chelates or complexes. Due to physical, chemical and biological reactions within the water, there may be dynamic interactions among the various particle sizes and chemical forms. Standard analytical techniques may not be able to detect the metals in all their particle sizes and combined forms, thus providing less than meaningful results. Consideration of the analytical techniques employed must be made in interpreting the results obtained.

There are several types of potential interactions that can take place between incoming metals and the water bodies they enter that will affect the concentration and distribution of the metals in the water. For example, the pH and Eh of the receiving water can control solubilization or agglomeration and therefore subsequent sedimentation of the metal species. The pH also affects the bonding of metals to insoluble carriers since hydrogen ions can influence adsorption and ion exchange by competing for active sites, modifying the sites or changing the degree of protolysis of the sorbing material. Morgan and Stumm [47] for example, showed that the sorption of Mn^{2+} on freshly precipitated MnO_2 was greatly influenced by pH due to changes in the sorption sites. Hydrous oxides and coprecipitated hydrous oxides and clay minerals generally exhibit strong cation exchange properties in alkaline solutions, strongly binding many hdyrolyzable metals to their surfaces. As the pH decreases, the surface charges and attractive forces become diminished, until finally in acid environments most materials become anion exchangers and will no longer hold positively charged metals, but will bond vanadates, chromates and similar complex metal ions with a negative charge [48].

The precipitation of a substance such as calcium carbonate through an increase in pH can remove metals like Zn and Cu through adsorption and coprecipitation. A change in the pH of a natural water can change the degree of complexation of a metal in solution since many complexing agents are also weak acids or bases. Usually an optimum pH or pH range exists for the stability of a given metal chelate or complex. In many cases the favored metal will displace other metals from the organic moiety and will be solubilized or insolubilized by the chelate or complex formation.

The oxidation-reduction potential, Eh, exerts a similar effect on certain transition metals and often supplements the effect of a pH change. Iron and manganese are the most responsive to Eh changes. Lower redox potentials favor the Fe^{2+} and Mn^{2+} valence states that are much more soluble than the oxidized states. Since Mn is more easily reducible than Fe, a significant amount of Mn^{2+} ion can accumulate in water having a slightly higher redox potential than that in which Fe^{2+} will be stable. Consequently, as redox potentials change in natural waters, there may be disproportionate changes in the concentration of Mn as compared to Fe. Vanadium, on the other hand, becomes more insoluble in the V(III) state as compared to the V(V), particularly at low pH. The higher, oxidized state is soluble over a wide pH range. Redox potential, like pH, will also affect the stability of certain transition metal chelates.

Further types of reactions are the formation of metal carbonates or sulfides having limited solubility. In some cases, complex ions may also result from these reactions leading to highly soluble forms of the metal. For example, at 75°C and 20 atm pressure (in H_2S saturated water), the solubility of ZnS is reported to be a million times higher than its solubility product would predict [5]. Hutchinson [49] attributes the increase in hypolimnetic copper to the diffusion into the water of metallo-organic compounds such as copper sulfide-amino complexes formed in the lake sediments. On the other hand, the addition of high concentrations of soluble copper to high carbonate waters, such as those resulting from copper sulfate treatment or the discharge of acid mine wastes, leads to rapid condensation and sedimentation of copper carbonate and subsequent low levels of copper in the overlying waters.

Other inorganic ions present in natural waters may also increase the solubilization of certain transition metals. For example, as chlorosity is increased, a greater proportion of manganese becomes dissolved rather than suspended. Lentsch, *et al.* [50] attributes the slight increase in total manganese with increasing chlorosity in part to leaching from the suspended material and, more importantly, to the leaching of manganese from the sediments. A similar effect has been noted when the runoff from salted roads enters fresh water streams and lakes. Kharkar, *et al.* [51] noted a similar increase in dissolved cobalt at the junction of fresh and saline waters. Chromium, however, did not show this pattern. Complexing and chelating reactions may include both inorganic and organic complex ion formation. Chau, *et al.* [52] concluded from ultrafiltration studies that most of the metal complexing compounds in lake water fall into an arbitrary molecular weight range of 1,000 to 10,000.

Two important metal complexing agents found in natural waters are humic and fulvic acids. These form stable metal humates and fulvates that are soluble in fresh waters. The stability of the complex will vary with the pH of the water. Bondarenko [53], for example, reports that Cu fulvates are more stable at moderately low pH in fresh water while humates are more stable at neutral and slightly alkaline pH. In sea water the presence of divalent cations causes precipitation of the metal fulvates and humates along with some inorganic phases.

Physical-chemical reactions involving adsorption and ion-exchange play important roles in binding metals to insoluble particles. These reactions are particularly important in flowing waters where significant amounts of clay and soil particles are entrained in the water. Kharkar, et al. [51] report finding large fractions of the cobalt sorbed on the clay particles in several rivers.

Another group of interactions of metals and receiving waters can be classed as biochemical, or reactions involving living organisms. Much of the dissolved and filterable organic matter that reacts with metals arises from aquatic organisms. For example, Lee and Hoadley [54] found that the concentration of dissolved organic carbon in Lake Mendota, Wisconsin, is about 10 mg/l, and the organic carbon in filterable particles is about 2 mg/l, mainly derived from excretory and degradation products of plants and animals. The biochemical processes of living organisms are also responsible for many of the changes in pH and Eh, which in turn affect metal concentrations. In addition to these indirect effects, living organisms play a direct role in controlling concentrations and distributions of metals in water bodies.

Almost all aquatic organisms concentrate some, if not all, of the transition metals in their tissues or skeletons. The concentration factors vary with the different metals and types of organisms as well as with the aqueous environment in which the assimilation is taking place. Table 2.21 lists some typical metal concentration factors shown by various fresh water and marine organisms. Martin [61] has summarized various mechanisms by which aquatic organisms assimilate metals as: (1) particulate ingestion of matter containing metals suspended in the water, (2) ingestion of food, (3) solubilization and assimilation through

Table 2.21

Typical Metal Concentration Factors of Selected Aquatic Organisms*

	Marine Organisms				Fresh Water Organisms		
Metal	Phytoplankton[a]	Zooplankton[b]	Macrophytes[b]	Mollusks[c]	Fish[d]	Macrophytes[e]	Mollusks[f]	Fish[e]
Ti	2,700	-	-	-	-	-	-	-
Cr	7,800	-	2,880	21,800	-	-	267	10
Mn	3,800	3,900	-	2,300	373	1,450	-	23
Fe	28,300	114,600	-	14,400	-	3,642	-	190
Ni	570	560	1,050	4,000	235	-	650	85
Co	-	-	-	-	50	1,367	300	90
Cu	2,800	1,800	2,890	3,800	127	158	1,500	60
Zn	5,500	8,800	7,000	27,300	533	318	2,258	228

* μg metal/g organic
μg metal/gram H$_2$O

[a]Martin and Knauer [55]
[b]Hägerhäll [56]
[c]Pringle, et al. [57]
[d]Goldberg [58]
[e]Merlini, et al. [59]
[f]Mathis and Cummings [60]

secretion of biological chelating or complexing agents, (4) incorporation into physiological systems, and (5) ion exchange and sorption on tissue and membrane surfaces.

The assimilation of metals by aquatic organisms will generally act to reduce their concentration in the waters surrounding the organisms and will increase the concentrations of the metals in the sediments that ultimately receive the biological material in which the metals have been concentrated. The assimilation of metals by aquatic organisms will also often result in higher metals concentrations at depths where organisms are more abundant or in the deeper waters of the water body, resulting from microbial breakdown of the biological material as it settles through the water column or from adsorption of settling material. Martin [61], for example, has found higher average values of Cu, Fe, Mn, Ni, and Zn in zooplankton samples from the deep oceanic waters and has concluded that more of these elements are associated with copepod exoskeletons as they settle to the sediments, due to adsorption processes. The secretion of chelating agents by some organisms can act to increase, rather than decrease as is most often the case, the concentration of soluble metals in the immediate vicinity of the organism. However, this process is usually highly localized.

DISTRIBUTION IN TYPICAL NATURAL WATERS

There is such a great variability in sources of metals available and entering a particular water and in the chemistry and/or biology of the water body that meaningful generalizations of metal concentrations in various types of natural waters, such as oligotrophic lakes and polluted rivers, are not possible. However, it is interesting to observe the concentrations and distribution of metals in a few typical natural waters to gain an appreciation of their occurrence and variability. Three types of natural waters will be considered, but it must be remembered that each blends into another with no true point of distinction since distinct "types" do not exist in nature. Discussed below are oligotrophic-mesotrophic mountain lakes with bedrock basins, mesotrophic-eutrophic lakes with glacial/fluvial till or limestone basins, and streams and rivers in both of the above basins.

Oligotrophic-Mesotrophic Lakes

These lakes are characterized by granitic and sedimentary glacial-scoured or volcanic rock basins, thin soil layers in the drainage basin with coniferous or mixed coniferous-deciduous forests, soft water with small amounts of humic substances, and only minor anthropogenic influence. Examples would be the precambrian Shield lakes of Canada, Lake Superior, Lake Tahoe, Crater Lake, and similar mountain lakes.

Lake George, New York, a glacial lake located at the eastern edge of the Adirondack Mountains, is representative of this type of lake. It lies in a glacial-scoured basin of precambrian metamorphic, plutonic and igneous rock with small patches of Cambrian deposits mainly at the southern end of the basin. Morainic deposits occur in many places around the lake and in the drainage basin. A shallow soil cover formed from glacial debris covers most of the drainage basin, with numerous rock outcrops occurring. Most of the watershed is covered by deciduous forest with numerous conifers also present. White pine, hemlock, hard maple, beech and yellow birch are the dominant tree species. Precipitation is the major water input to Lake George, with an annual precipitation of about 93 cm (37 in). About 0.1 km^3 of water falls directly on the lake each year, and 0.2 km^3 enters through many small streams. The drainage basin of the lake has an area of 606 km^2 with 19% of this area (114 km^2) covered by the lake.

The lake is long, slightly winding and oriented in a north-northeasterly direction with the outlet at the northern end. The lake was formed by the glacial scouring and damming of two preglacial river channels, giving rise to a north lake and a south lake separated by a relatively shallow island-studded midportion called the "Narrows." The south lake has been further divided into two basins, south and central, on the basis of morphometric and circulation characteristics [62,63]; each contains one very deep portion and several shallower subsidiary areas. The deep south basin, also called the Caldwell basin, is exposed to much greater anthropogenic influence in the form of motor boating, marina operation, and seepage from sewage treatment plants and septic tank effluents than is the north lake.

Periodic measurements of iron, manganese, copper and zinc, along with other chemical parameters, were made of the lake water as well as of the precipitation falling on the lake and in the influent streams during 1970-71 [30]. Table 2.22 lists the mean seasonal dissolved concentrations of these four metals in the north

Table 2.22

Mean Seasonal Concentrations of Fe, Mn, Cu and Zn
in the North and South Lakes of Lake George

Season	Depth (m)	South Lake (µg/l)				North Lake (µg/l)			
		Fe	Mn	Cu	Zn	Fe	Mn	Cu	Zn
Winter	3	27.2	2.0	5.2	43.4	35.2	1.9	2.7	51.1
(Jan. 1-Mar. 31)	9	42.1	2.1	3.5	49.3	34.8	1.3	2.0	79.6
	15	30.6	1.6	3.7	44.4	50.7	2.3	2.2	76.6
Spring	3	25.1	3.2	3.9	32.7	41.5	2.9	2.6	33.5
(Apr. 1-June 21)	9	17.3	2.5	4.2	28.0	26.2	2.5	3.5	53.2
	15	16.9	4.0	3.8	30.4	35.4	3.2	3.2	38.6
Summer	3	29.0	2.6	3.4	46.4	29.8	2.0	3.0	74.9
(June 21-Sept. 21)	9	23.5	2.2	3.1	31.8	23.8	3.3	3.2	40.4
	15	28.8	4.1	2.9	34.2	23.6	1.9	2.9	23.9
Fall	3	46.1	1.8	3.1	25.1	13.8	1.4	1.6	71.1
(Sept. 21-Dec. 7)	9	39.9	1.7	2.5	23.3	20.5	1.2	1.7	88.3
	15	30.3	2.5	2.6	43.5	14.5	1.1	2.0	74.5

and south lakes of Lake George at three depths. Table 2.23 lists the seasonal concentrations of these elements entering the lake in the precipitation and streams. The amount of iron, manganese and zinc in the suspended material in the lake (filter residue) was too low to be measured (< 5 µg/l for Fe and < 0.25 µg/l for Mn and Zn). Copper was occasionally as high as 1 µg/l in the suspended material but most of the time was less than 0.25 µg/l. The levels of dissolved iron, manganese and copper in the lake are consistently lower than the concentrations in the entering streams and precipitation, indicating that these elements are being lost to the sediments.

Table 2.23

Mean Concentrations of Fe, Mn, Cu and Zn in Lake George Influent Streams and Precipitation

	Streams (µg/l)				Precipitation (µg/l)			
	Fe	Mn	Cu	Zn	Fe	Mn	Cu	Zn
Winter	48.8	7.1	3.6	27.7	43.1	3.8	6.4	31.3
Spring	38.5	2.1	3.9	8.9	75.0	8.1	11.2	48.5
Summer	39.7	7.5	3.7	7.3	45.1	3.2	7.6	32.1
Fall	42.5	7.2	6.8	10.0	36.0	7.5	21.4	84.9
Snowpack on lake					6.2	5.6	23.8	23.0

The lakes of northeastern Minnesota are very similar to Lake George in that they lie in precambrian bedrock, have only a thin soil cover and are surrounded by mixed coniferous-deciduous forests. Most of them have less anthropogenic influence than Lake George. Bright [64] studied these lakes intensively. He reports mean dissolved metal concentrations in these lakes of 94.0 µg/l Fe, 16.8 µg/l Mn, 6.2 µg/l Cu and 11.6 µg/l Zn. The iron and manganese are considerably higher than found in Lake George and can be attributed to the higher levels of humic and tannic acids in the Minnesota lakes and the lower alkalinities as well as the larger accumulations of iron and manganese oxides in the Minnesota glacial deposits. The copper is slightly higher and the zinc lower than that found in Lake George.

Bradford, et al. [65] determined the concentrations of several metals in 170 lakes in the California High Sierras. The results demonstrate the great natural variability that can be encountered in different lakes of the same general type. Some of these lakes were evidently fed from geothermal sources of varying composition since they contained unusually high concentrations of certain metals such as molybdenum, vanadium and boron. The median concentrations of Fe, Mn, Cu and Zn are much lower in the Sierra lakes than in the Minnesota lakes or Lake George. The range measured by Bradford and associates encompasses the concentrations measured in Lake George.

Hagen and Langeland [66] report dissolved concentrations of 90 µg/l Fe, 15 µg/l Zn, and 10 µg/l Cu and a pH of 6.2 in Lake Fløvatn, an oligotrophic precambrian bedrock lake in southern Norway. This lake was receiving a larger input of metals and acid in the precipitation than Lake George. For example, the snow section taken from the snow pack on the Norwegian lakes contained average concentrations of 25 µg/l Fe, 10 µg/l Mn, 10 µg/l Cu and 30 µg/l Zn. The copper concentration is significantly higher in the Lake George snow than in the Norwegian snow. Measurements made of local grey zones in the Norwegian snow, indicative of high short-duration fallout, gave values of 15 and 20 µg/l copper, which are comparable to the Lake George snow, 35 µg/l Mn, 80-180 µg/l Fe and 85 µg/l of zinc. Lake Øvre Lomtjørn, also in Norway, would be classified as dysoligotrophic since it contains considerably more humic substances than a typical oligotrophic lake. This is the principal reason for the high iron observed in the lake, particularly in the deeper waters [66].

The effect of humic substances can be seen in the dysoligotrophic lakes of central Finland. These lands lie in precambrian granitic bedrock but are surrounded by numerous peat bogs and glacial drift containing more basic rocks such as amphibolites and gabbroes. Salo and Voipio [67] report mean concentrations of 500 µg/l Fe, 44 µg/l Mn, 33 µg/l Cu, and 76 µg/l Zn in Lake Kuolinjoki and surrounding streams. This lake has a catchment area of 95 km^2, of which the lake surface makes up to 10.4%. The pH of the lake averages 6.8.

The dissolved iron concentration of 40 Canadian Shield lakes, which lie in precambrian granitic basins with thin soil covers and coniferous or mixed coniferous-deciduous forests, ranged between 10 and 100 µg/l with a median concentration of 50 µg/l [68]. The pH range was 5.4-7.5. The concentration of iron in other soft water oligotrophic lakes is shown in Table 2.24. The iron content of Lake Superior shown in the table is very high (360 µg/l). If not due to nonrepresentative samples, it is felt this can be attributed to the large iron deposits adjoining the lake. Weiler and Chawla [70] report a mean concentration of 12 µg/l of copper in Lake Superior, with a range in concentration of 4-230 µg/l. They also report a mean concentration of 27 µg/l of zinc in the lake with a range of 9-80 µg/l.

Table 2.24

Iron Concentration in Oligotrophic Lakes*

Lake and Region	Total Fe (mg/l)	pH	TDS (mg/l)	Conduc. µmho/cm	No. Lakes
Canadian Shield					
Wollaston L., Sask.	0.005	7.0-7.3	35	35	1
Large lakes, NWT	0.01	-	10	20	7
NE Wisconsin	0.04	-	-	37	8-530
Cree L., Sask.	nil	6.8-7.0	27	-	1
Eagle L., Ont.	0.01	7.3	-	68	1
Lac Seul (at Ear Falls)	0.04	7.6	64	79	1
Kam L., NWT	0.07	-	125	-	1
L. Nipigon	0.10	7.8	108	149	1
L. Nipissing	-	7.4	59	76	1
Partial Shield Drainage					
Flin Flon area, Man.	0.01	-	185	242	16
Gt. Bear L., NWT	0.13	7.8	-	155	1
L. of the Woods	0.15	7.8	100	111	1
L. Superior	0.36	7.4	82	79	1
Sask. & Churchill drainage	0.024	8.1-8.2	136	-	5 large
Other Dilute Lake Districts					
Adirondack Mts., N.Y.	0.14	5.6	-	29	9
Chandler L., Alaska	0.04	-	47	-	-
Ikrowik L., Alaska	{0.01 / 0.05	-	36 / 85	-	-
L. Haruna, Japan	0.01	-	75	-	1
L. Yamanaka, Japan	0.05	-	74.3	-	-
L. Chujenji, Japan	0.08	-	100	-	-

*from Armstrong and Schlinder [68], and Livingston [69].

In summary, there is a relatively large variation in the metal concentrations in oligotrophic mountain lakes, depending upon the available sources and the amount of complex-forming humic substances. Because of the small amounts of sestonic material in many of these lakes, most of the metals present are in soluble or colloidal form rather than suspended. Since the waters are usually well oxygenated and of low productivity, there is little variation in metal concentration with depth as compared with mesotrophic and eutrophic lakes.

Mesotrophic-Eutrophic Lakes

The lakes in this grouping are characterized by glacial or fluvial till, clay or soil basins with some rock outcroppings, deciduous forest or cleared areas, prairies or plains adjoining the lake, medium to hard water with $\geqslant 200$ mg/l dissolved solids, and significant anthropogenic influence. Some examples are Lake Erie, Lake Ontario and the southern part of Lake Michigan of the Great Lakes, the Finger Lakes of New York, southern Lake Champlain, N.Y., Linsley Pond, Conn., Lake Mendota, Wis. and most of the lakes of the central and southern part of the United States.

Saratoga Lake, N.Y., lying approximately 20 miles south of Lake George, but in glacial and fluvial deposits overlying paleozoic rock strata, is typical of this type of lake. Saratoga Lake receives a relatively large amount of nutrient input and associated metals from nearby towns and lakeside cottages. The average hardness of the lake is 83 mg/l as $CaCO_3$, and the pH ranges from a minimum of 7.4 in the surface waters in the winter to a maximum of 8.7 in October in these same surface waters. Table 2.25 lists the average concentrations of Fe, Mn, Cu, Zn and hardness during the various seasons for Saratoga Lake epilimnetic and hypolimnetic waters. During the summer and fall, the main stream entering the lake, the Kayaderosseras, has a hardness of 96 mg/l as $CaCO_3$; the average soluble iron is less than 0.7 μg/l, manganese is 14.4 μg/l, copper is 1.5 μg/l, and zinc is less than 0.5 μg/l.

The table shows that there is a striking increase in hypolimnetic concentrations of suspended Fe and Mn in the fall. These samples were taken in October before the lake became isothermal. The buildup in hypolimnetic iron and manganese in the summer and fall is typical of lakes of this type, although the increases are usually seen in both the suspended and dissolved fractions. When H_2S is generated in the sediments in the summer and fall, such as occurs in Saratoga Lake, the solubility of the

Table 2.25

Mean Metal Concentrations and Hardness Found in Saratoga Lake

	Winter	Spring	Summer	Fall
		Epilimnion		
Fe (µg/l)				
Dissolved	28.4	12.5	23.7	34.0
Particulate	34.2	14.2	28.2	35
Mn (µg/l)				
Dissolved	4.5	2.9	3.4	4.9
Particulate	9.1	1.8	2.2	5.6
Cu (µg/l)				
Dissolved	10.8	2.9	2.2	3.5
Particulate	0.57	2.9	2.1	1.8
Zn (µg/l)				
Dissolved	29.6	<0.4	<0.8	<0.8
Particulate	0.57	15.0	<0.5	3.5
Hardness				
(mg/l as $CaCO_3$)	84	82	82	86
		Hypolimnion		
Fe (µg/l)				
Dissolved	22.4	11.6	15.1	36.0
Particulate	20	15.6	39.1	148
Mn (µg/l)				
Dissolved	2.0	1.8	2.8	4.8
Particulate	8.5	2.2	11.7	711
Cu (µg/l)				
Dissolved	7.0	3.5	3.7	2.6
Particulate	1.5	2.8	0.4	0.4
Zn (µg/l)				
Dissolved	11.5	<0.4	<0.4	<0.8
Particulate	0.5	10.6	<0.4	0.4
Hardness				
(mg/l as $CaCO_3$)	80	80	77	72

iron and manganese is greatly reduced at alkaline pH, and large increases are seen only in the particulate or total fraction. This same sort of behavior has been observed in Linsley Pond with iron and manganese, with hydrogen sulfide forming in the sediment [49].

In Saratoga Lake, the highest levels of particulate zinc occurred in the spring and the highest levels of dissolved zinc occurred in the winter throughout the water column. The dissolved zinc became undetectable throughout the lake from the spring on. The particulate zinc, after becoming nondetectable in the summer, reached measurable levels in the fall before isothermal conditions and also showed a decrease in depth. The copper concentrations and seasonal changes in Saratoga Lake are similar to those reported by Kimball [71] on Knight's Pond and Riley [72] on three dimictic ponds in Connecticut. The dissolved copper increases in late fall, reaching its highest levels during late fall and winter, then decreases throughout the late winter and spring, reaching the lowest values in the summer and early fall.

Par Pond in South Carolina illustrates the behavior of metals when sulfides are not formed. Par Pond, which could be classed as mesotrophic, is generally similar to southern deciduous forest and prairie lakes except that the concentrations of dissolved solids are extremely low and more typical of granitic basin mountain lakes. The lake has a much higher level of seston than a typical low solids mountain lake and in this parameter is similar to other mesotrophic-eutrophic lakes.

Marshall and LeRoy [73] found an increase in both dissolved and sestonic iron in the hypolimnion at the end of the summer. The dissolved manganese also increased in the hypolimnion but the particulate manganese was lower in the hypolimnion at the end of the summer than earlier in the year. The soluble manganese reached a maximum concentration of 2400 $\mu g/l$ at the end of one summer. The dissolved zinc did not show any large change in concentration with depth; however it did appear to have a slightly higher concentration in the hypolimnion than in the surface waters when two year means were considered. The highest dissolved zinc values generally occurred during the summer, although during 1966 high values in the surface waters were measured in the winter and spring. In Par Pond, most of the zinc was soluble. The particulate zinc, although low, showed both a seasonal and depth trend reaching its peak in the summer or later in the year and showing a slight decrease with depth. Kubota, *et al.* [74] also found that the highest dissolved zinc and copper concentrations occurred in June and July in Cayuga Lake, New York, decreasing in August. The mean and maximum concentrations of metals reported in this mesotrophic lake were, respectively, 2.17 and 9.41 µg/l Zn, 0.6 and 1.57 µg/l Cu, and 0.02 and 0.09 µg/l Co. These levels are lower than those found in most oligotrophic mountain lakes.

In general, where there is no strong anthropogenic introduction of metals, the surface waters of mesotrophic-eutrophic lakes in limestone or sedimentary basins, such as Cayuga Lake, have lower concentrations of the transition metals and zinc than do oligotrophic granitic basin lakes. Typical is Bright's [64] summary, as shown in Table 2.26, of the differences he found in the average dissolved concentrations of Fe, Mn, Cu and Zn in the surface waters of the granitic bedrock coniferous forest lakes of northeast Minnesota and the prairie and deciduous forest lakes of the other parts of the state.

Table 2.26

Average Dissolved Iron, Manganese, Copper and Zinc Concentrations in Typical Minnesota Lakes*

	(µg/l)			
	Fe	*Mn*	*Cu*	*Zn*
Coniferous granitic lakes	94	16.8	6.2	11.6
Deciduous forest lakes	14	2.0	3.6	7.3
Prairie lakes	7.3	2.3	3.7	3.7

*from Bright [64]

The introduction by man of metals into these medium-to-hard water lakes can cause relatively high local concentrations of metals in the lake water. Chau, *et al.* [52] give a good description of the variation in dissolved metal concentrations in Lake Ontario due to industrial pollution in the western and northeastern part of the lake. They found iron concentrations of 20 to 25 µg/l in industrial inshore portions of the lake compared with 1 to 8 µg/l in the central portion of the lake. Copper followed the iron with high concentrations of 10-14 µg/l occurring in the regions where high iron was found and 1-4 µg/l in the central part of the lake. Zinc was more variable than iron or copper but showed the highest values in the region of industrial development, *i.e.*, 50 µg/l in the western region and 20-30 µg/l in the eastern region. The zinc in most of the lake was below 5 µg/l. Manganese was generally low in the lake (<1 µg/l), which is common for mesotrophic-eutrophic lakes. The highest values (2-4 µg/l) were found in the western industrialized region offshore of Hamilton and Toronto. The nickel concentrations in the lake were generally between 1 and 3 µg/l, with the highest concentration of 8.5 µg/l near Hamilton. Similar high local concentrations have been seen in Lake Erie and in certain portions of Lake Michigan where the high levels have also been related to industrial pollution.

Streams and Rivers

As shown by Leopold [75], streams have their start in a series of parallel rills on a more or less even slope. Individual rills overflow and run into each other, forming a typical dendritic drainage pattern, ultimately leading to large streams and rivers. The water drained in this manner comes almost entirely from precipitation, either in the form of runoff on or through the soil or as shallow seepage. In some cases, it also comes from water that has percolated through the soil, from shallow seepage, and from water that has percolated through the earth to the water table, which in turn intersects and feeds the stream or river. As a consequence, streams will be greatly influenced by the rocks and soils of their drainage. The more the precipitation percolates through the earth before reaching the stream as compared to direct runoff over and through the soil, the more material there will be dissolved in the water and the less carried in suspension.

This distribution of precipitation reaching the streams is greatly influenced by the amount and kind of vegetation present as well as the thickness of the soil and earth cover overlying the bedrock. In streams running through forested areas, the vegetation will often have a greater effect on the amount of dissolved organic materials and metals in the water than in adjoining lakes because of the relatively small water volumes of the streams and their long shoreline exposed to leaf litter. Streams flowing through limestone regions will tend to pick up more metals from the rocks than would lakes situated in limestone basins because the stream will erode away the rock with its movement and the metals in the eroded rocks will enter the stream and be transported. Similarly, man's discharges to the streams can produce higher local concentrations of both soluble and particulate total metals in the stream than in a lake.

Streams and rivers can be divided into two main categories, fast-moving mountain streams in granitic basins having thin soil cover and coniferous or mixed coniferous-deciduous forests, and slower-moving streams in basins with thick earth cover, with urban grassland or deciduous shrubs and forest cover. This latter type of stream is more frequently subjected to extensive anthropogenic influence.

Over its 315 mile length, the Hudson River in New York has many regions typical of each of these classifications. The upper Hudson is a typical fast-flowing mountain stream lying in a precambrian bedrock extension of the Canadian Shield. The surrounding area is covered with only a thin soil layer supporting a mixed coniferous-deciduous forest. There is practically no habitation in this region (designated Region 1 in Table 2.27). Farther south at Luzerne and Corinth, N.Y., the river widens, decreases in velocity and flows through a region of glacial till and sedimentary rock with most of the land cleared. Small patches of deciduous forest may line the shore occasionally. Anthropogenic influence is present in the form of a few industrial and municipal discharges. This is designated as Region 2. At Troy, N.Y., a tidal dam is present; below the dam, the Hudson is tidal with a mean tidal range of 5.5 feet. It also receives a large quantity of both treated and untreated industrial and municipal waste along with the wastes brought in by the Mohawk River. The oxygen concentration is often near zero

Table 2.27

Typical Metal Concentrations in Various Regions of the Hudson River

	Fe		Cr		Mn		Ni		Cu		Zn			mg/l		
					μg/l											
	Dis.	Sus.	Dis.	Sus.	Dis.	Sus.	Dis.	Sus.	Dis.	Sus.	Dis.	Sus.	Alk.	Hard.	pH
Hudson River															
Region 1*	89.5	82.3	0.3	2.5	2.5	3.4	2.4	2.2	3.2	1.7	4.3	0.5	13.0	18	6.6
Region 2*	6.2	31.2	0.6	2.5	0.3	5.0	3.4	5.0	7.1	1.2	9.3	1.2	14.0	12	6.7
Region 3*	57.2	178.0	3.6	2.5	13.3	22.4	21.4	2.7	10.0	3.1	15.8	3.7	55.5	72	7.1
Region 4†	58.0	-	6.0	-	2.1	-	1.0	-	11.0	-	57.0	-	50.0	80	7.4
Streams influent to the Hudson															
Region 1*	15.5	6.2	0.3	2.5	1.8	0.6	3.7	2.2	3.4	0.5	0.9	0.5	10.0	14	6.4
Region 2*	34.2	16.7	0.3	2.5	1.2	2.0	6.3	1.3	3.6	0.5	0.6	0.5	18.0	24	7.0
Region 3*	16.7	68.8	9.6	2.5	17.4	22.5	86.8	1.9	13.6	1.9	5.8	1.1	214.0	493	7.7

*S. L. Williams, unpublished data.
†Public Health Service [76].

for about 30 miles south of Troy, with coliform counts from 1,000 to 50,000/100 ml in this portion of the stream, which is designated Region 3.

Table 2.27 lists the typical dissolved and suspended metal concentrations in these three regions of the Hudson along with other pertinent parameters. Typical concentrations are also listed for the river at Poughkeepsie, N.Y., about 80 miles south of Troy (Region 4). The stretch of the Hudson in the precambrian-coniferous region shows higher dissolved iron and manganese than farther downstream, where the river flows through glacial till and sedimentary rocks. Little change in the chemical nature of the water is evident but significant dilution from incoming streams has no doubt occurred. The incoming streams in Region 1 are high in manganese, probably from the forest litter because these measurements were made in the fall. The copper and zinc are higher in Region 2, probably as a result of municipal and industrial discharges since the unpolluted incoming streams show low levels of copper and zinc. In Regions 3 and 4, where large municipal and industrial discharges have occurred, there is an increase in hardness and alkalinity as well as in the concentrations of the dissolved metals. Soluble chromium is found in the river and influent streams in this region. Copper and zinc are also much higher.

In addition to the increase in dissolved metals in the polluted portions of the Hudson River, the table shows that a large increase in suspended iron and manganese occurs. This is due partly to the increase in pH, alkalinity and hardness of the river water, which causes some of the dissolved iron and manganese to precipitate. It is also due to the discharge of particulate matter to the river and the transport of suspended iron and manganese in the influent streams as indicated at the bottom of Table 2.27. The metal concentrations found in the various regions of the Hudson are typical of those found in similar streams and rivers.

By way of comparison, Turekian [77] offers the following as worldwide average concentrations of transition metals and Zn in natural streams (in μg/l): Sc - 0.004, Mn - 7, Cu - 7, Ti - 3, Zn - 20, V - 0.9, Co - 0.2, Cr - 1, and Ni - 0.3. Livingston [69] concludes that the mean concentration of Ni, Cu and Zn in fresh waters is 10 μg/l for each, with concentrations of the other metals varying too greatly to establish a mean concentration of much significance. Generally, dissolved iron is higher in soft water rivers whether polluted or unpolluted than in the hard water rivers, even those receiving metal wastes. Copper appears to show the same trend. The Mississippi at St. Paul and the Monongahela at Pittsburgh are two exceptions showing higher copper concentrations than many of the soft water polluted streams. The concentrations may vary with stream flow due to dilution. In unpolluted rivers, the nature of the underlying rock or soil plays an important role in the concentrations of dissolved metals by providing a source and by introducing ions such as carbonates and bicarbonates, which in turn will affect the solubility of the metals.

The Colorado River at Loma, Colorado, for example, contained 1.9-11.5 µg/l vanadium with a mean concentration of 4.6 µg/l which is attributed to natural weathering and solubilization of the vanadium-rich rocks in the basin [78]. The concentrations of the other metals, with the possible exception of zinc and iron, are low. The Allegheny and Monongahela also contain high vanadium, 7 µg/l and 4.3 µg/l, respectively [46]. This, however, is due to waste discharges from steelmaking operations.

Andelman [79] reviewed some of the factors affecting the concentration of metals in streams. He found that flow can have a varying effect on the metal concentrations, with increased flow causing either an increase or a decrease, depending upon the particular river system. Andelman also discussed the work of Masironi [80] who found positive correlations between Co, Cr, V, Ni and Zn and the hardness of the river water. Andelman attributes this correlation to the leaching of metals from the limestones and other Ca-Mg minerals exposed to the water. Durum and Haffty [81], in comparing large North American rivers, found that the ratio of the mean concentrations for Ni/Cr and Ni/Cu varied within a relatively small range, *i.e.*, 1.1-2.5 for Ni/Cr and 1.4-3.1 for Ni/Cu, indicating that chemical reactions in the river water, common to all of the rivers investigated, were controlling the relative concentrations of these elements. The lithologic environment could also be playing an important role in the relative availability of the metals, which in turn affected their concentrations.

SUMMARY

The three principal sources of transition metals and zinc entering the aquatic environment are the rocks, soil and plant litter of drainage basins; the dry and wet fallout of particles in the atmosphere; and the industrial and municipal discharges associated with man's activities. The concentration of metals in aquatic environments will depend not only upon the nature of their sources but also, and more importantly, on the chemical and biological nature of the receiving waters.

In waters having little or no anthropogenic influence, the low pH, low dissolved solids, oligotrophic mountain lakes and fast-flowing mountain streams generally contain higher concentrations of metals in their surface waters than do slower-moving waters of higher pH and higher concentrations of dissolved salts. The higher concentrations in mountain waters are usually due to the presence of greater quantities of humic substances to solubilize the metals and the absence of significant quantities of dissolved salts to react with and precipitate metal ions. In many cases, streams flowing through sedimentary regions, particularly limestone areas, will pick up greater quantities of metals through erosion and leaching of the limestone materials than will streams flowing over more resistant granitic rock.

In eutrophic and polluted waters, very high local concentrations of certain metals often occur as a result of an acid or strong reducing environment coupled with the introduction of the metals. The metals may come naturally from the sediments exposed to the reducing environment or from industrial or municipal discharges. While discharges of metals into high pH, high dissolved solids streams and lakes may cause high local concentrations of the metals, the metal concentrations soon decrease with distance from the polluting source due to the chemical and biological reactions taking place in the water.

REFERENCES

1. Turekian, K. K. and K. W. Wedepohl. *Bull. Geol. Soc. Am., 72,* 175 (1961).
2. Wager, L. R. *Observatory, 67,* 103 (1947).

3. Wager, L. R. and R. L. Mitchell. *Geochim et Cosmochim Acta, 1,* 129 (1951).
4. Riley, J. P. and G. Skirrow. *Chemical Oceanography,* Vol. 2. (New York: Academic Press, 1965).
5. Krejci-Graf, K. in *Encyclopedia of Geochemistry and Environmental Sciences,* R. W. Fairbridge, Ed. (New York: Van Nostrand-Reinhold 1972), p. 1201.
6. Whitehouse, A. E. and R. C. McCarter. in *Encyclopedia of Geochemistry and Environmental Sciences,* R. W. Fairbridge, Ed. (New York: Van Nostrand-Reinhold, 1959).
7. Krieger, R. A. Ph.D. Thesis, University of Minnesota (1949).
8. Zhukova, V. A. and E. A. Solodnikova. *Izu. Akad. Nauk. Kas. SSR Ser. Biol., 1,* 6 (1972).
9. Jenne, E. A. in *Trace Inorganics in Water.* (Washington, D.C.: American Chemical Society, Advances in Chemistry Series No. 73, 1968), p. 337.
10. Davies, B. E. *OIKOS, 22,* 366 (1971).
11. Little, P. and M. H. Martin. *Environ. Pollut., 3,* 241 (1972).
12. Nilsson, I. *OIKOS, 23,* 132 (1972).
13. Gosz, J. R., G. E. Likens and F. H. Borman. *Ecol. Monog., 43,* 173 (1973).
14. Battan, L. J. *The Unclean Sky.* (New York: Doubleday, 1966).
15. Wilson, C. P. *Man's Impact on the Global Environment.* (Boston: MIT Press, 1970).
16. Zoller, W. H., G. E. Gorden, E. S. Gladney and A. G. Jones. in *Trace Elements in the Environment.* (Washington, D.C.: American Chemical Society, Advances in Chemistry Series No. 123, 1973), p. 31.
17. National Air Pollution Control Administration. "Air Quality Data from the National Air Sampling Networks and Contributing State and Local Networks, 1964-65," Cincinnati, Ohio (1966).
18. American Chemical Society. "Cleaning Our Environment – The Chemical Basis of Action," Washington, D.C. (1969).
19. Zegger, J. L. Personal communication (1973).
20. Kluesener, J. W. Ph.D. Thesis. University of Wisconsin, Madison (1971).
21. Matheson, D. H. *Can. J. Technology, 29,* 406 (1951).
22. U.S. Public Health Service. "Air Pollution Measurements of the National Air Sampling Network—Analyses of Suspended Particulates, 1957-61," (Washington, D.C.: U.S. Govt. Printing Office, 1962).
23. Brar, S. S., D. M. Nelson, E. L. Kanabrocki, C. E. Moore, C. D. Burnham and D. M. Hattorie. in *Modern Trends in Activation Analysis,* J. R. DeVoe, Ed. (Washington, D.C.: National Bureau of Standards, Special Publication 312, 1969), p. 43.
24. Dudey, N. D., L. E. Ross, and V. E. Noshkin. in *Modern Trends in Activation Analysis,* J. R. DeVoe, Ed. (Washington, D.C.: National Bureau of Standards, Special Publication 312, 1969), p. 55.

25. Hoffman, G. L., R. A. Duce and W. H. Zoller. *Environ. Sci. Technol.,* 3, 1207 (1969).
26. Rancitelli, L. A., R. W. Perkins, T. M. Tanner, and C. W. Thomas. *Precipitation Scavenging* (Washington, D.C.: Atomic Energy Commission, Symposium Series 22, 1970).
27. Morrow, N. L. and R. S. Brief. *Environ. Sci. Technol.,* 5, 786 (1971).
28. Smith, R. G. in *Metallic Contaminants and Human Health,* D. H. K. Lee, Ed. (New York: Academic Press, 1972), p. 142.
29. Bressan, D. J., R. A. Com and P. E. Wilkniss. in *Advances in Chemistry,* Series 123. (Washington, D. C.: American Chemical Society, 1973).
30. Williams, S. L., E. M. Colon, R. Kohberger, and N. L. Clesceri. in *Bioassay Techniques and Environmental Chemistry,* G. E. Glass, Ed. (Ann Arbor, Mich: Ann Arbor Science Publishers, 1973).
31. Lazrus, A. L., E. Lorange, and J. P. Lodge, Jr. *Environ. Sci. Technol.,* 4, 55 (1970).
32. Rahn, K. A. Thesis, University of Michigan, Ann Arbor, Mich. (1971).
33. Zoller, W. H., E. S. Gladney and R. A. Duce. *Science, 183,* 198 (1974).
34. Fruh, G. *J. Water Pollut. Control Fed., 39,* 1449 (1967).
35. Allaway, W. H. *Advan. Agron., 20,* 235 (1968).
36. Patterson, J. B. E. *Technical Bulletin, Ministry of Agriculture, Fisheries and Food, 21.* (Cambridge, England: Agriculture Development and Advisory Service, 1971).
37. U.S. Dept. of the Interior. "Water Quality Criteria," Report of the National Technical Advisory Committee, Federal Water Pollution Control Administration, Washington, D.C. (1968).
38. U.S. Dept. of Health, Education and Welfare. "Drinking Water Standards," Revised, U.S. Govt. PHS Publication, Public Health Service, Washington, D.C. (1963).
39. National Academy of Sciences – National Academy of Engineering. "Water Quality Criteria 1972," U.S. Govt. Printing Office, Washington, D.C. (1973).
40. Klein, L. A., M. Lang, N. Nash, and S. L. Kirschner. "Sources of Metals in New York City Wastewater," Presented at the 46th Annual Meeting of the N.Y. Water Pollution Control Assoc., January 1974.
41. U.S. Public Health Service. "Interaction of Heavy Metals in Biological Sewage Treatment Processes," Public Health Service Publication 999-WP-22, Washington, D.C. (1965).
42. Kirkham, M. B. "Disposal of Sludge on Land; Effects on Soils, Plants and Ground Water," Presented at the 46th Annual Meeting of the N.Y. Water Pollution Control Assoc., January 1974.
43. Blakeslee, P. A. "Monitoring Considerations for Municipal Wastewater Effluent and Sludge Application to the Land," U.S. Environmental Protection Agency, U.S. Dept. of Agriculture, Universities Workshop, Champaign-Urbana, Illinois, July 1973.

44. Argo, D. G. and G. L. Culp. *Water and Sewage Works, 119*, 62 (1972).
45. Pitt, R. E. and G. Amy. *Environmental Protection Technology Series* (Washington, D.C.: U.S. Govt. Printing Office, EPA-R-2-73-283, 1973).
46. Kopp, J. F. and R. C. Kroner. *J. Water Pollut. Control Fed., 39*, 1659 (1967).
47. Morgan, J. J. and W. Stumm. *J. Colloid Sci., 19*, 347 (1964).
48. Park, G. A. in *Equilibrium Concepts in Natural Water Systems.* (Washington, D.C.: American Chemical Society, Advances in Chemistry Series 67, 1967), p. 121.
49. Hutchinson, G. E. *A Treatise in Limnology*, Vol. I. (New York: Wiley, 1957).
50. Lentsch, T. W., J. T. Kneip, M. E. Wrenn, G. P. Howells and M. Eisenbud. in *Radionuclides in Ecosystems*, D. J. Nelson, Ed. (Washington, D.C.: National Technical Information Systems, Conf. 710501-P2, 1971), p. 752.
51. Kharkar, D. P., K. K. Turekian and K. K. Bertine. *Geochim. Cosmochim. Acta, 32*, 285 (1968).
52. Chau, T. K., V. K. Chawla, H. F. Nicholson, and R. A. Vollenweider. *Proc. Great Lakes Res. Conf., 13*, 659 (1970).
53. Bondarenko, G. P. *Geokhimija,88*, 1012 (1972).
54. Lee, G. F. and A. W. Hoadley. in *Equilibrium Concepts in Natural Water Systems.* (Washington, D.C.: American Chemical Society, Advances in Chemistry Series 67, 1967), p. 319.
55. Martin, J. H. and G. A. Knauer. *Geochim. Cosmochim. Acta, 37*, 1639 (1973).
56. Hägerhäll, B. *Botanica Marina, 16*, 53 (1973).
57. Pringle, B. H., D. E. Hissong, E. L. Katz, and S. T. Mulawka. *J. San. Eng. Div., ASCE, SA3*, 455 (1968).
58. Goldberg, E. D. *Limnol. Oceanog., 7*, 72 (1962).
59. Merlini, M., C. Rigliocca, A. Berg, and G. Pozzi. "Trends in the Concentration of Heavy Metals in Organisms of a Mesotrophic Lake as Determined by Activation Analysis," IAEA/SM-142/27 (1970).
60. Mathis, J. and T. F. Cummings. *J. Water Pollut. Control Fed., 45*, 1573 (1973).
61. Martin, J. H. *Limnol. Oceanog., 15*, 756 (1970).
62. Langmuir, I., J. T. Scott, E. G. Walter, R. Stewart and W. X. Rozon. *Langmuir Circulations and Internal Waves in Lake George* (Albany, N.Y.: Atmospheric Sciences Research Center, State Univ. of New York, Publication No. 42, 1966).
63. Needham, J. G., C. Juday, E. Moore, C. E. Sibley, and J. W. Titcomb. "A Biological Survey of Lake George, N.Y." New York Conserv. Comm. (1922).
64. Bright, R. C. "Surface Water Chemistry of Some Minnesota Lakes with Preliminary Notes on Diatoms," Bell Museum of Natural History, Univ. of Minnesota (1968).
65. Bradford, G. R., F. L. Bair and V. Hunsaker. *Limnol. Oceanog., 13*, 526 (1968).

66. Hagen, A. and A. Langeland. *Environ. Pollut., 5,* 45 (1973).
67. Salo, A. and A. Voipio. "Radioactive Contamination of the Marine Environment," IAEA Symposium, Volume STI/pub., Vienna (1973).
68. Armstrong, F. A. J. and D. W. Schindler. *J. Fish. Res. Bd., Canada, 28,* 1 (1971).
69. Livingston, D. A. "Chemical Composition of Rivers and Lakes," U.S. Geological Survey Prof. Paper 440-G, Washington, D.C. (1963).
70. Weiler, R. R. and V. K. Chawla. *Proc. Gr. Lakes Res. Conf., 12,* 801 (1969).
71. Kimball, K. D. *Limnol. Oceanog., 18,* 169 (1973).
72. Riley, J. P. *Ecol. Monog., 9,* 53 (1939).
73. Marshall, J. S. and J. H. LeRoy. in *Radionuclides in Ecosystems,* D. J. Nelson, Ed. (Washington, D.C., Conf. 710501-P2, NTIS, 1971).
74. Kubota, J., E. L. Mills and R. T. Oglesby, *Environ. Sci. Technol., 8,* 243 (1974).
75. Leopold, L. B. *Fluvial Processes in Geomorphology.* (San Francisco: Freeman, 1964), p. 12.
76. U.S. Public Health Service. *Pollution of the Hudson River and Its Tributaries.* (Washington, D.C.: U.S. Dept. of Health, Education and Welfare, 1965).
77. Turekian, K. K. in *Yearbook of Science and Technology.* (New York: McGraw-Hill, 1969).
78. Linstedt, K. D. and P. Kruger. in *Modern Trends in Activation Analysis,* J. R. DeVoe, Ed. (Washington, D.C.: National Bureau of Standards, Special Publ. 312, 1970), p. 87.
79. Andelman, J. B. in *Trace Metals and Metal-Organic Interactions in Natural Water,* P. C. Singer, Ed. (Ann Arbor, Mich.: Ann Arbor Science Publishers, 1973), p. 57.
80. Masironi, R. *Bull. World Health Org., 43,* 687 (1970).
81. Durum, W. H. and J. Haffty. *Geochim. Cosmochim. Acta, 27,* 1 (1963).

CHAPTER 3

LEAD AND MERCURY IN THE AQUATIC ENVIRONMENT OF WESTERN WASHINGTON STATE

W. R. Schell and R. S. Barnes
Laboratory of Radiation Ecology
College of Fisheries
University of Washington
Seattle, Washington

Biogeochemical Cycling	130
Experimental Methods	132
Sample Collection of Sediment, Biota and Water	132
Lead Procedure	134
Mercury Procedure	135
Sediment Method	135
Results and Discussion	138
Sediment Studies – Mercury	138
Sediment Studies – Lead	140
Geochemical Budget and Pollution Vectors	143
Water	144
Biota	153
Conclusions and Recommendations	162
Summary	163

The program reported here is part of a general study to establish base line values for the amounts of lead and mercury in the biota and water from Lake Washington, Puget Sound and tidal waters of Washington State. The sources of these heavy metals in the aquatic environment are fallout, erosion, stream runoff, leaching from tidelands and sediment, as well as additional amounts from industrialization and urbanization of the region. In order to make the analytical data relevant to the problem of the concentration of heavy metals in the food chain leading to man, it is important to understand the pathways through the trophic levels. It is also important to have a means of separating natural sources from manmade sources during the time history of "industrial man" in this local region. From data on the past build-up of pollutants, it may be possible to evaluate the present rationally as well as to estimate future expected concentrations. This paper summarizes the analytical and collection techniques developed for lead and mercury and the results of analyses on the bioenvironmental evaluation of the concentrations found in samples.

BIOGEOCHEMICAL CYCLING

The processes of transport, distribution, and removal to sediments are part of the geochemical cycle. Pollutant metals are additional quantities of elements superimposed on the normal geochemical cycle. It is important to assess the magnitude of the sources produced by man and his technology that are imposed on the natural metals cycle in the geochemical system of atmosphere, hydrosphere, and lithosphere. An assessment of the past concentrations and the increase over the time of industrial man in a given region are necessary to gain perspective of the impact of pollution on the ecological system.

The most significant modes of trace metals transport are shown in Table 3.1, where the amounts of heavy metals produced and the potential transport to the ocean are given. The removal of trace metals to a geochemical sink must include an aqueous phase. Removal processes can include adsorption, precipitation, oxidation, reduction and complex formation. These processes rapidly remove many trace metals from the aquatic environment and deposit them in sediment within a short distance of the input [1]. It is for this reason, coupled with large dilution factors, that the concentrations of many trace elements in water are low. However, if the trace metals are changed in chemical form in the sediments by oxidation-reduction, bacterial action, or organometal complex formation, they may become incorporated into the biota and transferred up the food chain to man.

Table 3.1

Global Trace Metal Production and Potential Ocean Input*

Element	Mining Production (10^6 tons/yr)	Transport by Rivers to Oceans (10^6 tons/yr)	Atmospheric Washout (10^6 tons/yr)	Ratio Aeolian Fluvial
Pb	3	0.1	0.3	3.0
Cu	6	0.25	0.2	0.8
V	0.02	0.03	0.02	0.7
Ni	0.5	0.01	0.03	3.0
Cr	2	0.04	0.02	0.5
Sn	0.2	0.002	0.03	15
Hg	0.009	0.003	0.08	27
Cd	0.01	0.0005	0.01	20

*from National Academy of Sciences [2]

Advective transport is the most important mode of redistribution of trace metals from the lithosphere. The several modes identified are: (1) those that are carried by the wind (aeolian), (2) those that are carried by running water, either suspended or dissolved (fluvial), and, (3) those that are removed or transported by rain or other forms of precipitation scavenging (pluvial). Aeolian transport has been noted since the twelfth century [3]. Meridional and zonal dust transport have been observed from the Sahara to the mid-European continent [4], from Africa to the Caribbean [5], and throughout the world from the volcano Krakatoa. Additionally, such dust transport has resulted from radioactive fallout originating at Eniwetok and Bikini, Marshall Islands, Novya Zemlya, USSR, Lop Nor, and China. These sources have served as tracers of geochemical processes and have permitted calculations of residence times in the atmosphere to be made [6-9]. The recent findings on the build-up of lead and mercury in Greenland ice profiles illustrate long range aeolian transport [10]. Short range aeolian transport is noted in the high concentrations of trace metals in the air of industrial North American cities [11,12].

The problems arising from local pollution sources of trace metals may be transported thousands of kilometers from the point of origin. Thus, pollution from industrialized nations becomes an international concern [13]. It is clear that effective control of potential trace metal pollutants can be exercised only at the source where the quantity and chemical form can be regulated.

EXPERIMENTAL METHODS

For this study representative samples in the food web were obtained, and specific chemical separation and analytical procedures for the analyses were developed.

Sample Collection of Sediment, Biota and Water

In order to follow the pathway of heavy metals from their source to man, the following collections were made:

1. water: soluble (inorganic-organic on filter bed collection—LVWS); insoluble (particulate >0.3-μ Millipore filter collection—LVWS)
2. plankton: nannoplankton, zooplankton (copepods, crustaceans, larva, etc.)
3. pelagic fishes: herring, smelt, salmon
4. benthic fishes: sole, sablefish, cod
5. intertidal organisms: barnacles, mussels, oysters, clams, crabs
6 sediment.

The collections of water, sediment and organisms were made using the research vessel *Commando*. Figure 3.1 shows the location of sampling stations. The biota were sampled with a one-meter "0" mesh plankton net, a 10-foot Issacs Kidd midwater trawl and an otter trawl. The sediment samples were taken using a one-inch-diameter gravity corer and a piston corer. Water samples were obtained at various depths using a large volume water sampler (LVWS) [14], including a one-inch polyvinyl chloride tubing and pump [15]. The nannoplankton and detritus were collected on 0.3-μ Millipore filters while the soluble components were collected on chromatographic grade Al_2O_3. Immediately after collection the samples were frozen for later analysis.

Lead and Mercury in the Aquatic Environment 133

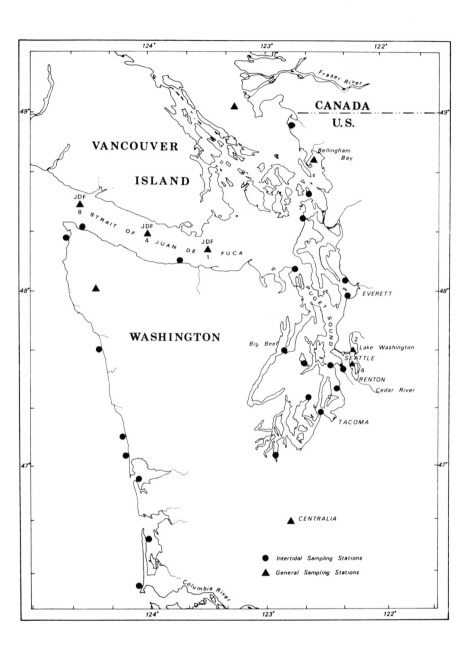

Figure 3.1. Collection stations in western Washington.

Lead Procedure

Biological samples for stable lead are either wet-ashed with a concentrated HNO_3-$HClO_4$ mixture, or dry-ashed after partial dissolution with concentrated HNO_3. The choice of ashing methods depended partly on the additional elements to be analyzed in the same sample. In either case the samples were spiked with ^{212}Pb, which had been separated from a natural thorium salt, to measure the chemical yield [16]. After ashing, the sample was dissolved in 0.3 M HCl and extracted twice with 10 ml of 15-g/l DDTC solution (diethyldithiocarbamic acid, diethylammonium salt) in chloroform. The extract was evaporated to dryness with a few milliliters of concentrated HNO_3, and charred; heating in the muffle furnace at 550°C for one hour followed. The residue was then taken up in 0.3 M HCl and diluted to volume. The chemical yield of lead was determined by counting the ^{212}Pb tracer using a 3 x 3 NaI (Tl) crystal, phototube and scaler. Total Pb in the sample was determined using a Perkin-Elmer Model 303 atomic absorption spectrometer. The reagents HNO_3, HCl and water used were distilled to reduce lead contamination. A reagent blank was run with each batch of eight samples to determine the total lead contamination from external sources.

The sorption beds from the LVWS were treated in a slightly different manner due to the different matrix, Al_2O_3 [15]. The entire Al_2O_3 or an aliquot was first dried and placed in a 50-mm diameter glass column and spiked with ^{212}Pb tracer. Lead was then eluted with a large volume (500 ml) of 8 M HNO_3 and 6 M HCl. The resultant eluate was evaporated to about 200 ml and the organic compounds removed by wet-ashing with concentrated $HClO_4$. Lead yield was then determined by counting ^{212}Pb and total lead determined on the atomic absorption spectrometer as before. A matrix and reagent blank were determined with unused Al_2O_3 from the appropriate lot. The lead values of the blank ranged from about 75 to 200 µg total, depending on the lot used.

Millipore filters from the LVWS were treated like biological materials but were wet-ashed because the filters were cellulose nitrate and would burn violently if muffled. A typical value of total lead for these filters was about 0.7 µg per filter. Filter diameter was 300 mm, and a maximum of eight filters was used per sample.

Mercury Procedure

The cold vapor (flameless) atomic absorption technique was used for the determination of mercury in water, sediments, and biological material. Poluektov, *et al.* [17] were among the first to use this technique for detection of trace quantities of mercury. Hatch and Ott [18] refined the method and extended its sensitivity to submicrogram levels. Further improvement in the sensitivity of the method to nanogram levels has resulted primarily from better instrumentation. Much of the recent mercury methodology has been directed toward adapting existing techniques to a wide variety of sample materials having low natural mercury concentrations. The method of Melton, *et al.* [19], modified by the addition of a hot digestion step, provides a simple, sensitive and rapid technique for mercury determinations in a wide variety of materials. The Melton technique has been used routinely for water, sediments, and biological samples.

The mercury procedure, using cold vapor (flameless) atomic absorption (FAA),* has been checked against a number of interlaboratory standards as well as by neutron activation (NA). The results are shown in Table 3.2. Interlaboratory standards were measured over a period of 18 months and the largest relative error on these standards has been 16 per cent for the NBS orchard leaves. Over a one-month period replicate analysis of groups of three identical samples of the NBS tuna meal gave values that agree within 5 per cent.

Sediment Method

The evaluation of contemporary transport and distribution of trace metals in a given region can be achieved by systematic collection and analysis of air and water samples, dry fallout,

---

*Mercometer Model 2006-1, Anti-Pollution Technology Corporation, Holland, Michigan, with a 75-cm absorption cell.

Table 3.2
Comparison of Standard Samples with Other Laboratories and Methods

	LRE (FAA) ($\mu g/g \pm \sigma$)	Other Labs ($\mu g/g \pm \sigma$)
Sediment		
Homogenized Bellingham Bay sediment	1.66 ± 0.04	1.70 ± 0.06[a]
		1.72 ± 0.13[b]
Homogenized dry sediment		
EPA 72C5643	106.5 ± 2.9	109.55 ± 21.50[c]
72C5644	41.8 ± 0.95	43.91 ± 9.13[c]
72C5645	0.0619 ± 0.0033	0.22 ± 0.38[c]
Biological Materials		
NBS orchard leaves	0.121 ± 0.003	0.140 ± 0.015[d]
NBS bovine liver	0.0143 ± 0.0043	0.0160 ± 0.002
NBS tuna meal	0.792 ± 0.032	0.60 to 0.99[e]
Water		
Local interlaboratory	0.0093 ± 0.0003	0.0093
Local interlaboratory	0.0052 ± 0.0002	0.0050
EPA #3 organic and inorganic Hg (3 ml)	0.0064 ± 0.0002	0.0063[f]
EPA #4 organic Hg (5 ml)	0.004 ± 0.0001	0.0042[f]

[a]Comparison with U of W oceanography laboratory - FAA
[b]Comparison with U of W oceanography laboratory - NA
[c]Comparison with EPA mercury in sediment round-robin June 1972
[d]Tentative value
[e]Range of results reported in sample is due to inhomogeneity according to NBS
[f]Determined at the 20 ng level after dilution

precipitation, soils and sediments. While this comprehensive technique is satisfactory for evaluating contemporary events it cannot determine the history of trace metals in a given region. To determine the history of trace metal build up in a given region, soil and sediment profiles are most suitable. Because of weathering, natural mixing, and disruption by human activities, undisturbed soil profiles are difficult to obtain. However, sediment cores are often uniquely suited for the preservation of trace metals history.

Sediment cores for trace metals have the following advantages:

1. they are relatively independent from weathering.
2. Sedimentary deposits, in general, increase at a fairly constant rate per year.
3. Sedimentary material scavenges trace metals.
4. They are usually outside the sphere of continual human disruption but frequently contain anthropogenic anomalies that can be specifically dated.

Two basic assumptions are necessary in using sediment cores for the trace metal history. First, the system must be closed to gains and losses of the trace metal at all levels in the core, and second, a sequential accumulation must have taken place. Since the anthropogenic increase in trace elements has occurred in recent years, it is essential to have time resolution to within a few years. Practically, this demands a sedimentation rate of greater than two mm/yr.

The application of sediment cores to the time history of trace metals in Lake Washington has been made. Figure 3.1 shows the Lake Washington area and the major drainage basin, the Cedar River. The urban and suburban development of the region has been increasing rapidly since 1940. Lake Washington has an area of 88 km^2 and a mean depth of 33 m. Seattle lies along the western shore with suburban development surrounding the lake. It is spanned by two floating bridges, one completed in 1941, the other in 1963. A ship canal and locks were built in 1916, connecting the lake with Lake Union and Puget Sound. This lowered the lake some 3 m to its current elevation of 4.3 m above mean sea level. To maintain sufficient flow of water to operate the locks, the Cedar River, with an average flow of

19 m^3/sec, was permanently diverted into the southern end of the lake in 1916.

The lake has a history of cultural eutrophication, but little direct industrial waste discharge is believed to have occurred. Recovery from cultural eutrophication has been rapid since 1968 [20]. On the lower Cedar River, the Boeing Company, the Pacific Car and Foundry, and the city of Renton are potential major sources of trace metals to the lake.

RESULTS AND DISCUSSION

Sediment Studies - Mercury

The locations in Lake Washington of core sampling for trace metals are shown in Figure 3.1. The mercury profiles for the three cores taken are shown in Figure 3.2. The two cores from Station 2 show similar structure; however, the absolute mercury concentration is different. A fixed point in time has been found at the 18-cm level at Station 2 and at 25-cm at Station 4, where a silt band was produced when the lake level was lowered three meters and the Cedar River diversion and Lake Washington Ship Canal were completed in 1916. On this basis the sediment rate was found to be 3.3 mm/yr for Station 2 [21], in agreement with reported values at a nearby location [22]. For Station 4, by fitting the similar mercury structure and the silt layer, a sedimentation rate of 4.4 mm/yr is indicated. This is consistent with the core location that was closer to the Cedar River, a major source of sediment material.

The discrepancy between stations is in the absolute magnitude of the mercury concentration. The similarity in profiles indicates that general lake conditions are being reflected. It is apparent that the mercury associated with the greater sedimentation rate has a lower concentration; however, the total mercury available to the sediment could be constant. The dilution by low mercury sediment could account for the discrepancy between the cores from Station 4 and Station 2. The source of mercury to the sediments must be dispersed throughout the lake, indicating that the mercury in the lake is added at about the same rate

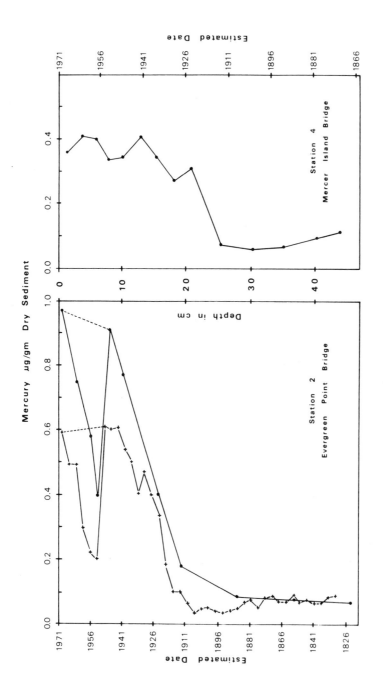

Figure 3.2. Mercury in Lake Washington cores.

throughout. The dramatic decrease at 4-6 cm for lead and mercury at Station 2 represents what is believed to be redistribution of silt due to the construction of the second floating bridge, completed in 1963. The difference between the 1963 data and the estimated date shown in Figure 3.2 of about 1956 is due to the higher water content in the upper few centimeters of the core. These core section samples had high silica content in contrast to the other samples. The X-ray radiographs of cores taken nearby show a density anomaly at the same depth, which could be attributed to the bridge construction [22]. Recent work by Edmondson (personal communication) indicates that diatom blooms were unusually heavy at this time (1963) and the resulting biogenic silicates could give the same dilution effect. Additional cores were taken in an attempt to define local and areal distribution of the mercury input.

At the present time it is not possible to sort the several suspected sources contributing to the mercury budget of the lake. Suspected sources for mercury include: combustion of fossil fuels, especially coal; aeolian and aeolian-fluvial transport from the Tacoma smelter; ·municipal sewage or treated effluents; and local industrial or domestic sources such as mercury in marine paints, and fungicides. The mercury associated with sewage wastes and storm runoff should be diminishing at present following the diversion of sewage from the lake in 1968.

Sediment Studies – Lead

The lead profile is shown in Figure 3.3 for a core taken from Station 2. The gross structure of lead concentration is similar to the mercury core, with the increase starting at the 18-cm level. The lead content is shown to increase by a factor of approximately 16 from the base line prior to 1916. The present sediment concentration of about 300 $\mu g/g$ indicates a continuous and increasing input source. The doubling time for the increase in lead concentration is about 18 years.

The fine structure of the core can be related to specific events by assuming the sedimentation rate of 3.3 mm/yr and no migration of the lead after deposition. The initial increase from base line at 18-20-cm depth can be ascribed to the diversion of

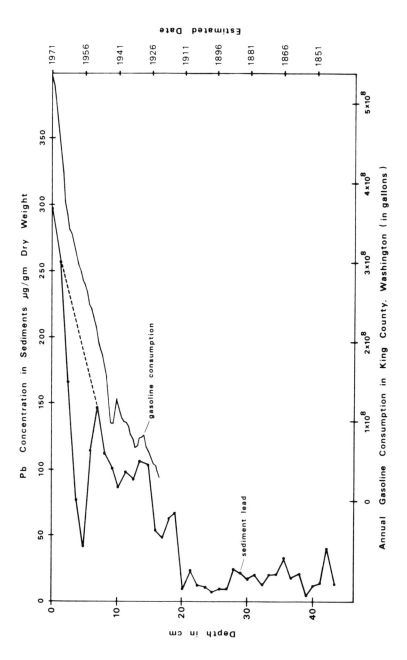

Figure 3.3. Lead in Lake Washington cores compared with gasoline consumption.

the Cedar River. The Cedar River drainage basin is about 500 km^2 in area and is generally downwind (E-NE) of a potential source, the American Smelting and Refining Company smelter in Tacoma. This smelter has been in operation since 1890 and has processed ore, mainly for lead, from 1890-1913. Stack emissions in 1971 may have been about 40% lead oxide [23]. Aeolian transport would bring lead to the drainage basin and then into the lake by fluvial transport via the Cedar River. The second rapid increase at the 14-15-cm level would correspond to the 1920's when lead alkyls were introduced in gasoline. The decrease in the lead deposition from the 14-19-cm levels could correspond to the depression and war years. The use of gasoline engines, predominately in the automobile, may be related directly to the lead profile. The 9-10-cm level may represent the postwar period when automobile use increased rapidly, resulting in a corresponding increase in lead concentration.

Data from the Washington State Department of Highways on gasoline consumption in this region was compared with the lead profile in Lake Washington sediments in Figure 3.3. The close similarity of these two curves, especially the reduction of gasoline consumption during the war years and the lower lead levels in the core coincident with this reduction, strongly implies a cause and effect relationship. This contention is also borne out by the doubling rate of gasoline consumption (15-16 years) and the nearly identical increase in lead build-up in the core (about 18 years). Agreement over the last decade is expected from the proximity of the core location to the Evergreen Point Bridge. The agreement during earlier years may indicate that more general environmental conditions are being reflected. Further core studies should better define local and areal lead distributions and their related causes.

The iron and sodium concentration profiles also have been measured. Since these elements are major constituents of the weathered materials, the concentration as a function of time (depth) indicates differences caused by erosional changes in the watershed, which drains into the lake. The profile shows fluctuations around a mean value of about 40 mg/g iron and 1.2 mg/g sodium. In contrast to trends of the other trace metals measured, there is no apparent long-term trend in the concentrations of sodium and iron in the sediment data [21]. The minimum at the 6-cm level is again evident and is for both metals. There appears to be a slight increase at the 16-18-cm level at the time the lake level was lowered and the Cedar River diversion was completed.

Geochemical Budget and Pollution Vectors

One way of approaching the pollutant input calculations for the budget is to consider the nature of the trace metal enrichment in the sediments since 1916. The enrichment in the contemporary surface sediment relative to the pre-1916 sediment has been 16 times for lead and 10 times for mercury. Table 3.3 shows the contemporary aeolian input of heavy metals to the sediments as estimated from the enrichment of these metals in the present sediment, assuming that the enrichment since 1916 is due entirely to aeolian material. This estimate is compared to that for aeolian input as derived from the net transport budget, which is a second way of approaching pollution input [21]. The results of both methods of calculation indicate an aeolian component greater than 90 per cent.

Table 3.3

Heavy Metals Input to Lake Washington Sediments

Element	Increase 1971/ pre 1916	Contemporary Aeolian Input to Sediment	
		Enrichment[a]	Budget[b]
Pb	16.6	94%	99%
Zn	3.9	74%	68%
Cu	3.6	72%	56%
Hg	10.0	91%	93%

[a]Assuming that all the heavy metal enrichment since 1916 is due to aeolian transport.

[b]On a net transport basis (see references 11,21,24).

Based on lake volume, fluvial output, and sedimentary deposition, the mean residence times were estimated as follows: iron, 11 days; lead, 26 days; mercury, 15 months; and sodium, 29 months [21]. The mean residence time for the lake water is 29 months. Such figures must be considered approximations since the assumption of complete mixing time, being much less than the mean residence time, does not appear valid during periods of stratification (June-October), at least for iron and lead.

There is evidence that the trace metals are primarily transported in the suspended load, often in inert positions within the suspended particles [25]. A further observation is that much of the detrital trace metals never leave the solid phase from initial rock weathering to detrital deposition. Lead entering the lake from the atmosphere appears to have a very short residence time in the water column, as shown by the low concentrations measured in the water in relation to the high deposition in the sediment. It is suspected that the same is true for other heavy metals in atmospheric material.

Table 3.4 shows the mercury concentrations in surface sediments collected in the Puget Sound region. Surface sediments from the Lake Washington drainage show low values in Findley Lake, the alpine headwaters of the Cedar River; intermediate values in Chester Morse Lake, an impoundment on the Cedar River within the closed watershed of Seattle; and considerably higher values in the metropolitan lowland lakes Sammamish, Washington and Union. Lake Union is part of the Lake Washington Ship Canal in the heart of Seattle and has been the site of considerable industrial and commercial activity for over 90 years. The sediment samples from Puget Sound near Seattle show little evidence of mercury pollution but they were not collected adjacent to industrialized areas. The effects of local industrial pollution are most apparent in the Bellingham Bay sediments at 3.95 $\mu g/g$ where the natural background is believed to be about 0.1 to 0.3 $\mu g/g$ [26].

Water

The concentrations of lead in the Lake Washington water samples taken with the LVWS are shown in Figure 3.4. The

Table 3.4

Mercury in Surface Sediments[a]

Location	Concentration ($\mu g/g$ dry)
Lake Washington drainage	
Findley Lake	0.048 (2)[b]
Chester Morse Lake	0.262 (2)
Lake Sammamish	0.435 (2)
Lake Washington	
North of Mercer Island Bridge	0.359 (1)
North of Evergreen Point Bridge	0.775 (2)
Off Sandpoint	0.704 (2)
Lake Union	
North end near old gas plant	1.18 (2)
Lake Union shipyard	2.47 (2)
Old sewer outfall, east side Lake Union	2.80 (2)
East side Lake Union	1.95 (2)
Puget Sound near Seattle	
West Point (86 m)	0.093 (2)
Old sewer outfall, West Point (27 m)	0.200 (2)
Northern Puget Sound	
Bellingham Bay off chloroalkali plant and pulpmill	3.95 (6)

[a] The analytical error is ± 10 per cent.
[b] Number of samples in parenthesis.

particulate fraction $>0.3\mu$ shows the seasonal spring maximum with minimum in summer to late fall. The soluble fraction shows some structure that is not easily explained. It is possible that use of the lake by gasoline boats contributes to the variable amounts of the soluble lead. The efficiency, E, for the soluble metal collected by the LVWS is calculated by $E = 1 - B/A$, where

146 Aqueous-Environmental Chemistry of Metals

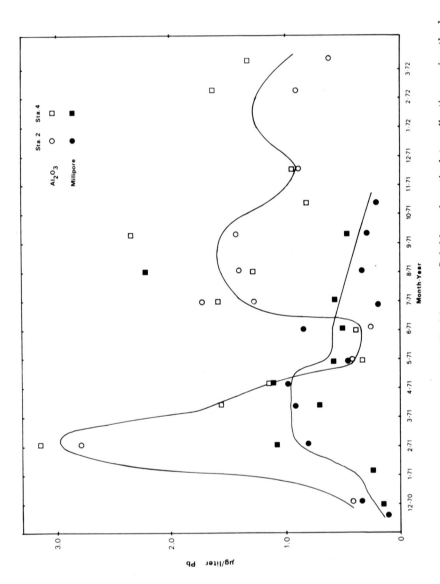

Figure 3.4. Lead concentration with time in Lake Washington. Soluble and particulate collections using the large volume water sampler.

A and B are the metal concentrations measured in the first and second Al_2O_3 beds [15,27]. The average total lead concentration in the lake water is about 2 µg/l.

The concentrations of lead at the "low pollution" Hood Canal salt water station (Big Beef) are shown in Figure 4.5. Here

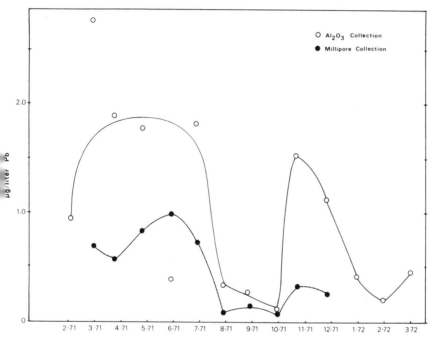

Figure 3.5. Lead concentration with time in Big Beef. Soluble and particulate collections using the large volume water sampler.

the particulate and soluble lead illustrate some different processes occurring in salt water. There appears to be a general increase in the lead concentration in spring in the soluble fraction with an abrupt decrease in July. The particulate concentration reaches a maximum in June and decreases abruptly to a minimum in August. These limited data give only trends that need to be confirmed by additional sampling. Certain samples may be contaminated by leaded gasoline from boats. The chemical state may change during the year due to environmental factors and different sources and should be measured. Thus, in deriving concentration factors for organisms for heavy metals it is essential to evaluate a time series of data on the water concentrations since short time variations can be large.

The lead concentration measured at salt water stations using the LVWS is shown in Table 3.5. The soluble fraction has been corrected for efficiency. Total lead concentration varied from 0.65-4.7 µg/l in the Juan de Fuca stations (JDF), with the higher concentrations found inland and in the deeper collections. This would be expected since ocean water enters the Strait of Juan de Fuca at depth and Puget Sound water exits at the surface. At Station JDF-1 the deeper collection was higher in both the soluble and particulate fractions than in the surface collection. This station is located near a sill where lower Puget Sound water mixes with the deeper Strait of Juan de Fuca water. At JDF-8 the deeper water had a lower concentration, indicating that low lead ocean water from depth dilutes the lead concentration. Tideland collections near Seattle showed the highest lead concentrations.

To compare the lead collection technique with other methods, samples from Amchitka Island, Aleutian Chain, were measured (Table 3.5). The particulate lead found was 0.06-0.09 µg/l and the soluble lead was 0.06-0.09 µg/l, giving a total lead concentration of 0.12-0.18 µg/l. These samples should be low and equivalent to surface ocean measurements reported by Chow and Patterson [28], who found that the lead concentration in surface oceans varied between 0.07-0.37 µg/l. The lead concentration technique using the LVWS did not appear to contaminate the sea water samples appreciably during collection or analysis.

The mercury concentration measured in water shows some interesting trends. The mercury concentrations in Lake Washington waters varied from 0.008-0.068 µg/l (Table 3.6) while the sediment concentrations gave 0.4-0.8 µg/g (Figure 3.2). These data indicate that the total mercury concentration in Lake Washington waters increases with depth, possibly suggesting that most of the mercury is associated with sinking detrital material or that the sediment is serving as both a sink and a source for mercury. The mercury concentrations in particulate fractions measured using the LVWS show that less than 10% of the total mercury is present as particulate (> 0.3 µ, 0.00054 µg/l).

Mercury concentrations for river waters collected November, 1971, in the vicinity of the new Centralia Power Plant placed in operation in the fall of 1971 (consumption about 8,000 tons of

Table 3.5

Lead Concentration in Salt Water Samples Using the Large Volume Water Sampler*

Station	Date	Depth	Volume Filtered (liters)	Soluble (μg/l)	Particulate (μg/l)	Total (μg/l)
North Seattle	30 June '71	Surface	2706	0.45	0.25	0.70
Duwamish	28 Apr. '71	Surface	1839	1.50	3.90	5.40
Mukilteo	17 Jan. '71	Tideland	441	1.95	3.23	5.18
JDF-1	25 Jan. '71	Surface	1900	0.49	1.07	1.56
JDF-1	25 Jan. '71	25 m	1938	1.90	2.83	4.73
JDF-4	26 Jan. '71	Surface	2432	0.82	1.36	2.18
JDF-4	26 Jan. '71	25 m	2166	1.26	0.92	2.28
JDF-8	26 Jan. '71	Surface	1900	0.74	0.24	0.98
JDF-8	26 Jan. '71	25-30 m	2090	0.37	0.28	0.65
Amchitka Island	9 Nov. '71	Tideland	566	0.09	0.09	0.18
Amchitka Island	20 Apr. '71	Tideland	589	0.06	0.06	0.12

*The analytical error is ± 10%.

Table 3.6

Mercury Levels in Lake Washington Waters
October-November 1971*

Location	Concentration ($\mu g/l$)
South End	
Surface	0.008
Bottom (26 m)	0.068
Cedar River	0.047
North of Mercer Island Bridge (Station 4)	
Surface	0.014
20 m	0.014
40 m	0.051
Madison Park	
Surface	0.008
20 m	0.008
40 m	0.023
60 m	0.045
North of Evergreen Point Bridge (Station 2)	
Surface	0.015
20 m	0.012
40 m	0.026
60 m	0.031

*The analytical error is ± 10%.

coal per day at peak operation), were in the range of 0.005-0.042 $\mu g/l$. One higher value of 0.116 $\mu g/l$ was reported just below the steam plant on Hanaford Creek; the concentration in the creek above the plant was 0.017 $\mu g/l$. Mercury concentrations seemed quite variable with the mean value for 18 water samples being 0.017 $\mu g/l$, excluding the steam plant value.

The mercury concentrations measured in salt water, as shown in Table 3.7, indicate that there was variability in the soluble collection efficiency at different locations using the LVWS sampler. These data indicate that different chemical states of mercury exist, or that the Al_2O_3 bed influenced chemical states during collection. The depth profile off the mouth of the Fraser

Table 3.7

Mercury Concentration in Salt Water Samples Using the Large Volume Water Sampler

Station	Date	Depth (m)	Vol.)(l)	Salinity (%)	Soluble Bed 1 (ng/l)[b]	Soluble Bed 2 (ng/l)	Eff.	Total (ng/l)	Insol. (ng/l)	Total (ng/l)
Fraser River	23 Mar. '72	Surface	696	19.989	2.60	0.83	0.68	3.80	1.93	5.73
"	"	15	988	29.797	1.43	0.40	0.72	2.00	0.85	2.85
"	"	35	969	29.918	1.48	2.17	?	?	0.91	?
Bellingham Bay	"	Surface	193	23.262	12.64	7.25	0.43	29.7	13.6	43.3
"	"	28	1056	29.567	1.54	2.26	?	?	0.98	?
JDF-1	20 Mar. '72	Surface	1090	29.946	1.58	1.54	?	?	1.06	?
"	"	65-74	1109	30.876	6.48	1.35	0.79	8.17	1.19	9.36
"	"	95-132	647	32.192	2.41	2.74	?	?	1.35	?
JDF-4	21 Mar. '72	Surface	1381	28.216	1.59	1.77	?	?	0.57	?
"	"	35-51	1582	30.500	1.38	1.27	?	?	0.42	?
"	"	75-128	776	33.352	1.79	2.22	?	?	0.87	?
JDF-8	"	Surface	1351	28.434	1.24	1.36	?	?	0.28	?
"	"	38	1575	31.020	1.38	1.11	0.20	7.07	0.48	7.55
"	"	107-140	988	33.069	2.16	1.63	0.30	7.34	0.56	7.90
Big Beef	31 May '72	Tidal	214	-	0.56	0.05	1.00	0.56	1.02	1.56
	15 Jun. '72	Tidal	413	-	0.17	0.02	1.00	0.17	3.65	3.82

[a]The analytical error is ± 10%.
[b]ng/l = 10^{-9} grams per liter.

River illustrates possible differences in chemical state of the mercury collected. The Al_2O_3 beds gave reasonable collection efficiency for the soluble mercury present (68%). The total mercury (soluble plus particulate) in the surface water was 5.73 ng/l. At mid-depth the salinity increased and the particulate mercury decreased. The Al_2O_3 beds gave similar soluble collection efficiency (72%), giving a total mercury content of about one-half the surface water, or 2.85 ng/l. Near the bottom the salinity increased only slightly and the particulate mercury remained almost the same at 0.91 ng/l. However, the second Al_2O_3 bed collected at a slightly higher rate than the first, probably due to different chemical states of mercury near the bottom. The organic mercury that may be produced would not be collected as efficiently by the Al_2O_3 beds as inorganic forms of mercury. The benthos may be serving as a source for mercury in the water column in a different chemical state than that being transported by the natural geochemical cycle from the river.

In Bellingham Bay the highest particulate mercury concentration was found in surface water (13.6 ng/l). The lower salinity indicates fresh water input at the sampling station, which is located at the outlet near the mouth of the bay where the bay water meets Puget Sound water. The Al_2O_3 bed collection efficiency was reasonable at 43%, giving a soluble mercury concentration of 29.7 ng/l. The total mercury at this point was 43.4 ng/l, the highest measured at any location sampled. The sampling station where the effluent from the bay would be sampled was located at a point far from the chloroalkali plant and the pulpmill; thus, concentrations should reflect low values since significant dilution by sound water would already have occurred. The particulate mercury content in the deep sample was similar to that at the Fraser River; the soluble mercury found on the Al_2O_3 beds was also similar in that it was not collected with measurable efficiency. As noted previously, the different chemical states of mercury found near the bottom may be responsible.

The Strait of Juan de Fuca sampling stations gave generally higher mercury concentrations inland than toward the sea. The low collection efficiency of the Al_2O_3 beds found in the Fraser River and Bellingham Bay was also observed in the strait. A trend toward increasing mercury concentration in the particulate fraction with depth is evident. The Al_2O_3 collection was not satisfactory to give estimates of the soluble fraction at all stations. Repeat sampling to better define systematic processes occurring in the particulate and soluble fractions would be desirable.

It appears that the mercury content of Puget Sound waters sampled, with a concentration range between 0.003-0.01 µg/kg, is systematically lower than that observed in many other areas of the world. From 70-90% of the mercury is found in the soluble fraction, *i.e.* less than 0.3 µ. This is in contrast to values in the Thames and its estuary where the mercury was found in the particulate fraction 80-98% [29]. In the Eastern Pacific near Mexico the total mercury concentration varied from 0.022-0.173 µg/kg [30]. In the English Channel the concentrations measured were from 0.010-0.21 µg/kg [31]. Recent measurements by Bothner and Carpenter (unpublished data) show total mercury concentrations of 0.01-0.02 µg/kg near the Washington coast and mouth of the Columbia River. The mercury pollution concentration of 0.043 µg/kg in the effluent of Bellingham Bay water shows up as a factor of about 10 above the natural level of mercury in Puget Sound but still below many measurements made in the open ocean.

Biota

The concentrations of lead and mercury in the sediments and water may be reflected in the biota sampled. A large number of sockeye salmon were collected and analyzed for lead over a two-year period. These fish were sized and the results reported for the different size fractions. Though an attempt was made to sample several size fractions each month from the two lake stations, this was not always successful due to the distribution pattern of fish in the lake. Results indicate no difference in the average lead concentration of the three sizes of juvenile sockeye collected, 50-75 mm, 95-120 mm, and 120-170 mm. The concentration in eviscerated whole fish averaged 0.54, 0.52, and 0.53 µg/g (dry), respectively, with a wet to dry ratio of 3.96. These averages were made from a total of about 50 samples of pooled collections consisting of 3-20 fish each. Thus, the juvenile sockeye in Lake Washington have lead concentrations of 0.13 µg/g (wet).

One of the most interesting findings of the studies is on a long-lived predator of the sockeye salmon, the northern squawfish (*Ptychocheilus oregonensis*). Figure 3.6 shows that the mercury content in muscle increases with age of the fish. This same trend has been found with northern pike in Sweden and in other long-lived fish [32,33]. Figure 3.6 also shows the lead

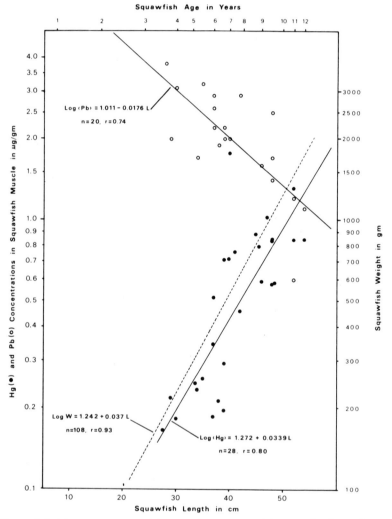

Figure 3.6. Concentration of lead and mercury with growth of northern squawfish.

concentration in these same fish. The decrease in concentration with age is apparent. In the analysis, the lead and mercury have an associated error of ± 10%.

The combined weight increase with length and the heavy metals concentration change with length of northern squawfish [34] is also shown in Figure 3.6. There is an apparent exponential increase in weight, W, in kg with increase in length, L, in cm following the equation fitted by least squares regression

$$\log W = 1.242 + 0.037 \, L \qquad (1)$$

The correlation coefficient of the data is 0.93.

The mercury concentration in mg/kg wet weight with length, L, in cm follows an exponential increase with length according to the equation

$$\log [Hg] = 1.272 + 0.034 \, L \qquad (2)$$

The correlation coefficient of the data is 0.80. The slopes of Equations 1 and 2 are similar, indicating that the mercury accumulation in muscle follows directly the increase in the mass of the northern squawfish. The lead concentration in mg/kg decreases with length, L, in cm according to the equation

$$\log [Pb] = 1.011 - 0.018 \, L \qquad (3)$$

The correlation coefficient of the data is 0.74. The ratio of the slopes of Equations 2 and 3 to the slope of Equation 1 illustrates the concentrating or eliminating properties of the organism and has the values +1.1 for mercury and -2.1 for lead.

Thus, it can be said that mercury concentrates in the muscle of the northern squawfish depending on the size (age) of the fish. The lead that is more concentrated in the young fish is eliminated and/or diluted as the fish become larger (older). If the initial lead absorbed in the muscle remained fixed and was diluted by the increase in mass of muscle of the growing fish, the slopes of Equations 2 and 3 would be equal but opposite in sign. However, it is shown that there is approximately a 50% difference in magnitude; the lead decrease (elimination) is slower than the weight increase. The reason for the significant differences in the heavy metals must be examined.

One possible explanation for the higher lead concentration of of the young northern squawfish may be found in the diet and feeding pattern. According to Olney [34] the diet of the young fish (3.8-7.2 cm) consists of diptera larvae (42.9%), unidentified insects (22.9%), cladocera (14.3%), diptera pupae (11.4%), odonata (2.9%), and amphipoda (2.9%). The larger fish, 22.5-62.0 cm, are predators whose diet consists of salmonids, smelt, and cottids. These food fish contain 0.06-0.4 μg/g Pb (wet).

It would appear that the insect larva could contain high concentrations of lead derived from the sediment, which could then be incorporated into the juvenile northern squawfish. Alternatively, physiological and/or biochemical changes, which may result in lower retention of lead, could occur as the fish mature. The early life history of these fish shows them to be concentrated mainly in the littoral zone where significant insect population exists. The high concentrations of lead in the juvenile squawfish may indicate that a source of lead is present in the benthos, is incorporated in insects and larva living in these areas, and is transferred by this pathway through the food chain. As shown, the lead concentration in the water column, though low, has a high concentration in the sediment, resulting in the short residence time calculated to be about 26 days. However, the short residence time may not remove the pollutant from the biological system.

Zooplankton are collected by towing a fine mesh net (1 m), a method that occasionally collects chips of paint and detritus as well. These will give exceptionally high lead concentrations. Such samples must be rejected when attempting to establish concentration ratios, because the zooplankton will have high lead concentrations. The zooplankton are primarily copepods and are the main food for the juvenile sockeye salmon and long finned smelt. Occasionally the viscera of the dissected fish contained exceptionally high lead concentrations, also indicating the presence of lead-containing debris. The concentration in zooplankton varied from 13-60 μg/g Pb (dry) with an average of about 24 μg/g Pb (dry). The wet/dry ratio can be assumed to be about 12.

From measurements of the water and biota the concentration ratios shown in Table 3.8 have been determined. As stated by

Table 3.8
Concentration Ratios for Various Organisms in Lake Washington and Puget Sound

Sample Type	Location	Size (cm)	Water (μg/kg)	Concentration (μg/g wet)	Concentration Ratio
Mercury					
Northern squawfish, muscle	Lake Washington	30	0.020	0.17	8,500
Northern squawfish, muscle	Lake Washington	50	0.020	0.80	40,000
Sockeye, muscle	Lake Washington	9.0	0.020	0.035	1,750
Longfin smelt, muscle	Lake Washington	9.3	0.020	0.067	3,400
Zooplankton	Lake Washington	-	0.020	0.013	650
Lead					
Northern squawfish, muscle	Lake Washington	30	2	3.4	1,700
Northern squawfish, muscle	Lake Washington	50	2	1.3	650
Sockeye, muscle	Lake Washington	5-17	2	0.13	65
Longfin smelt, muscle	Lake Washington	4-12	2	0.15	75
Zooplankton	Lake Washington	-	2	1-6	1,500
Oysters	Big Beef		1.3	0.10	75
Mussels	Big Beef		1.3	0.20	150

Lowman, et al. [35], concentration factors (ratios) are indicators
and not absolute since they may be altered by biological and
environmental factors. To illustrate that age is a parameter in
the concentration ratios of organisms, the evaluation of lead and
mercury concentration ratios have been determined. According to
Polikarpof [36], the concentration factor, CF, is equal to

$$CF = \frac{\text{concentration of metal in organism (wet weight)}}{\text{concentration of metal in water}}$$

Mercury follows the established pattern of increase in concentration with increased size of northern squawfish. The concentration ratio for this organism for mercury must be defined for different sizes. The mechanism of mercury transport through the food chain is not known; thus trophic level concentration ratios cannot be determined yet. The water concentration is 0.25-0.85 ng/l in the particulate fraction and about 8-50 ng/l total in unfiltered water. Thus, the average concentration of about 20 ng/l that is mainly soluble can be assumed and the concentration ratio calculated for young and old fish.

The lead concentration ratio can be calculated in the same manner. The water contains about 0.3-0.9 μg/l lead in the particulate fraction and about 0.5-2.0 μg/l in the soluble fraction depending on the time of year. The average total concentration can be assumed to be about 2 μg/l and the concentration ratio calculated for young and old fish.

An evaluation of the lead concentration in various regions of Puget Sound and western Washington State would be a monumental task if water concentrations were measured at representative stations even monthly over the year. The approach taken in these studies has been to follow the water concentrations and an indicator organism sampled monthly from one given "low pollution" region of the sound, namely, Big Beef station on Hood Canal. The organisms used were the molluscs: oyster (*Crassostrea gigas*) and mussel (*Mytilus edulus*). The systematic data on osyters were taken from the low pollution region; it was assumed that the concentrations of organisms from other regions could be compared directly if the concentration ratio could be determined at Big Beef station. By inference, the average lead concentration in water could then be estimated for organisms living in other areas. As shown in Figure 3.5, the lead concentration in the soluble and particulate fraction of the water varied significantly at Big Beef station over the years measured. However, for application to the concentration ratio, the average water concentration was about 1.3 µg/l. The lead concentration in the *C. gigas* collected over the year varied from 0.62-1.16 µg/g (dry) with an average of about 0.82 µg/g (dry) and a wet/dry ratio of 8. Thus, the concentration of 0.10 µg/g (wet) for *C. gigas* gives a concentration ratio for lead of 75.

The mussels (*Mytilus edulus*) were small in Big Beef station and a pooled collection of 10-50 was necessary for a sample, as indicated in Table 3.9. Only a few samples were measured; thus the concentration ratio of about 150 for lead in mussels is approximate. Since the oyster *C. gigas* is found only in certain areas and mussels are widely distributed, the mussel data has been used as the indicator of high lead concentrations. Table 3.9 shows representative data from various regions. The highest lead concentration, 4.5 µg/g (wet), was found in mussels growing at the Duwamish station. This station is in the center of the major population density of western Washington at the mouth of the river (see Figure 3.1). Other stations where high values were found were near the urban-industrial development, between Tacoma and Everett, where concentrations of 0.35-0.96 µg/g (wet) were measured. Though the Neah Bay station, where 0.7 µg/g (wet) was measured, is far from urban-industrial development, this region is a major small boat marine harbor where large quantities of leaded gasoline are used. The remaining stations contained 0.08-0.2 µg/g (wet) which may be assumed to be background lead concentration in mussels. Assuming a concentration factor of 150 for *M. edulus* as well as for *M. californianus*, the water concentration would be about 30 µg/l at the Duwamish station, 2.3-6.4 µg/l near the metropolitan Tacoma-Seattle-Everett area, and a natural background of 0.5-1.5 µg/l. The organism samples integrate the water concentrations over a significant time period, whereas water samples measure only single points in time. A point check is the station at Mukilteo where 5.2 µg/l was measured on 17 January 1972 (Table 3.5) and 5.3 µg/l was calculated from the mussel concentration.

These data can be compared with coastal waters of England where mean values at Liverpool Bay, Cardigan Bay, and Britsol

Table 3.9
Lead Concentration in Mussel Soft Parts Collected in Tidelands of Western Washington*

Station	Date	Species	Pb Concentration μg/g (dry)	Pb Concentration μg/g (wet)	Comments
Big Beef	22 Apr. '71	M. edulus	1.26	0.16	Low pollution stations for background
Seabeck	22 Apr. '71	M. edulus	1.95	0.28	
Seabeck	15 Jun. '72	M. edulus	2.44	0.21	
Olympia	9 Jun. '72	M. edulus	1.41	0.14	
Point Defiance	20 Mar. '72	M. edulus	2.56	0.24	Tacoma-Seattle-Everett population center
Point Defiance	10 Jun. '72	M. edulus	4.43	0.45	
Seahurst Park	24 Apr. '71	M. edulus	5.43	0.80	
Duwamish	14 Jun. '72	M. edulus	45.42	4.5	
Alki Point	2 Apr. '71	M. edulus	44.6	4.4	
Alki Point	14 Jun. '72	M. edulus	8.46	0.96	
Mukilteo	28 Mar. '71	M. edulus	13.4	0.80	
Mukilteo	13 Jun. '72	M. edulus	3.86	0.47	
Mission Beach	13 Jun. '72	M. edulus	1.43	0.13	
Deception Pass	26 Mar. '71	M. edulus	2.21	0.26	
Anacortes	26 Mar. '71	M. edulus	2.55	0.30	
Anacortes	11 Jun. '72	M. edulus	1.75	0.17	
Birch Bay	11 Jun. '72	M. edulus	1.06	0.16	
Agate Beach	22 Mar. '71	M. californianus	2.01	0.34	
Agate Beach	22 Mar. '71	M. californianus	1.18	0.18	
Agate Beach	29 Apr. '72	M. californianus	2.11	0.24	
Neah Bay	22 Mar. '71	M. californianus	3.31	0.69	
Neah Bay	22 Mar. '71	M. californianus	3.83	0.71	

Neah Bay	29 May '72	M. californianus	1.81	0.34
Makah Bay	22 Mar. '71	M. californianus	0.07	0.02
Makah Bay	22 Mar. '71	M. californianus	0.36	0.08
Makah Bay	29 May '72	M. californianus	0.96	0.18
Kalaloch	24 Mar. '71	M. californianus	0.95	0.16
Kalaloch	24 Mar. '71	M. californianus	0.89	0.14
Copalis	24 Mar. '71	M. californianus	0.75	0.13
Copalis	24 Mar. '71	M. californianus	0.72	0.12
Westport	23 Mar. '71	M. californianus	1.15	0.19
North Head	23 Mar. '71	M. californianus	2.11	0.31	Mouth of Columbia River

*The analytical error is ± 10%.

Channel were 1.74, 2.24, and 1.18 µg/l, respectively [1], with dissolved lead in the openoocean of 0.07-0.35 µg/kg [28].

Puget Sound waters contain lead above the natural levels originating primarily from the metropolitan areas. Since dilution is great by the tidal and circulation system of Puget Sound, the concentrations away from urban areas are low. However, with the increase in input with time, significant accumulation should be found in regions where there is increase in industrial-urban development as well as increase in the use of leaded gasoline in automobiles and boats.

CONCLUSIONS AND RECOMMENDATIONS

The natural geochemical cycle deposits heavy metals in sediments, with the majority being removed from the biological cycle. Mercury is concentrated in certain long-lived organisms, from a water concentration of nanograms per liter with the concentration ratio proportional to the age. Lead is concentrated in young northern squawfish and then is partially eliminated with age. This apparent elimination may be due to the diet of the fish, which changes with age from a primarily insect larva for the young fish to the predatory habits of the older fish. In order to evaluate the impact of heavy metals pollutants on an ecosystem, it is essential to study the transport of the trace metals through the trophic levels. Different organisms may concentrate or eliminate such toxins through a variety of pathways. The lead concentration of the sockeye salmon and long-finned smelt population in the lake is low even though the sediments contain 300 µg/g (dry). The concentration factor for lead in sockeye is low (about 65), and thus at the present concentration of water (2 µg/l) the concentration in wet muscle is only about 0.1 µg/l, *i.e.* below the "safe" level for human consumption. However, other food organisms, particularly benthic, may have different concentration mechanisms and should be investigated. The lead concentration in mussels growing near urban-industrial areas was excessive, indicating that the water concentrations were also high.

Even in such a low pollution region as the Pacific Northwest, the concentrations of trace metals in specific regions may exceed the permissible concentrations. Thus, further effort needs to be made in the area of determining the chemical state and ultimate fate of metal pollutants through the food chain. If these waters are used for aquaculture, these factors must be considered. Soluble *vs.* particulate concentrations of metals give indications of possible chemical state changes in various regions. However, the toxicity of different chemical states and organometallic complexes that may be formed in the biological system may be high, and even at the low concentrations present may be harmful to the organisms and ultimately to man.

SUMMARY

In summary, a systematic sampling and analysis program has been established for western Washington State. The geochemical cycling of trace metal pollutants of Lake Washington indicates primarily aeolian transport to the lake and its watershed. The mercury pollution as measured in the lake sediments does not appear to have increased significantly since the mid-1940's. The lead concentration is increasing rapidly with a doubling time of about 18 years, which appears to be correlated with gasoline consumption in automobiles. The lake water concentrations are low in mercury (6-68 ng/l). The lead concentrations in lake water (0.2-4.0 µg/l) show seasonal variation and possible contamination by leaded gasoline used in boats. The lead concentration in salt water at a low pollution station, Big Beef (0.1-3.0 µg/l) also shows seasonal variations. The average concentration in water available to biota for concentration must be evaluated by a time sequence sampling.

The mercury concentrations in Puget Sound (except for Bellingham Bay, a site where major mercury contamination is occurring) are systematically lower by a factor of 2-3 than many reported surface ocean concentrations. The biota in Lake Washington show concentration and elimination processes for mercury and lead. Northern squawfish show mercury concentration factors during growth from 8,500 to 40,000 for young and old fish, respectively. In contrast, lead is concentrated in young fish and the concentration decreases with age. Lead is accumulated in mussels growing near urban-industrial areas, so mussels serve as indicator organisms for lead contamination.

ACKNOWLEDGMENTS

Appreciation is expressed for the field collection and sample preparation by Messrs. Rodney Eagle, Arthur Johnson, and Fred Olney, and the crew of the *Commando*; for the construction of field collection equipment by Mr. Raymond Lusk; for the analytical efforts of Mr. Terrence Jokela and Mr. Edward Maxwell; and for the computer assistance on the data by Mr. Charles Vick and Mrs. Marguerite McAlpin. The primary funding for the program has been under EPA Grant No. R 800357 with additional support by AEC Contract No. AT(45-1)-2225-T14.

REFERENCES

1. Abdullah, M. I., L. G. Royle, and A. W. Morris. *Nature, 235,* 158 (1972).
2. National Academy of Sciences. *Marine Environmental Quality: Suggested Research Programs for Understanding Man's Effect on the Oceans.* (Washington, D.C.: National Academy of Sciences, 1971).
3. Rex, R. W. and E. D. Goldberg. in *The Sea*, M. N. Hill, Ed. (New York: Wiley-Interscience, 1962).
4. Walther, J. *Naturwiss., 18,* 603 (1903).
5. Brown, W. F. *Monthly Weather Rev., 80,* 59 (1952).
6. Fabian, P., W. F. Libby, and C. E. Palmer. *J. Geophys. Res., 73,* 3611 (1968).
7. Krey, P. W. and B. Krajewski. *J. Geophys. Res., 75,* 2901 (1970).
8. Pierson, D. H. and R. S. Cambray. *Nature, 216,* 755 (1967).
9. Schell, W. R., G. Sauzay, and B. R. Payne. *J. Geophys. Res., 75,* 2251 (1970).
10. Murozumi, M., T. J. Chow, and C. Patterson. *Geochim. Cosmochim. Acta, 33,* 1347 (1969).
11. Brar, S. S., D. M. Nelson, J. R. Kline, P. F. Gustafson, E. L. Kanabrocki, C. E. Moore, and D. M. Hattori. *J. Geophys. Res., 75,* 2939 (1970).
12. Lee, J. and R. E. Jervis. *Trans. Amer. Nucl. Soc., 11,* 50 (1968).
13. Bolin, B. *Proc. U.N. Conf. on the Human Environ., Sweden's Case Study.* (1971).
14. Silker, W. B., R. W. Perkins, and H. C. Rieck. *Ocean Engr., 2,* 49 (1971).
15. Schell, W. R., T. Jokela, and R. Eagle. IAEA/SM-158/47, IAEA, Vienna (1972).
16. Sill, C. W. and C. E. Willis. *Anal. Chem., 37,* 1176 (1965).
17. Poluektov, N. S., R. A. Vitkun, and Y. V. Zelyukova. *Zh. Anal. Khim, 19,* 937 (1964).
18. Hatch, W. R. and W. L. Ott. *Anal. Chem., 40,* 2035 (1968).
19. Melton, J. R., W. L. Hoover, and P. A. Howard. *Proc. Soil Sci. Soc. Amer., 35,* 850 (1971).

20. Edmondson, W. T. *Science, 169,* 690 (1970).
21. Barnes, R. S. and W. R. Schell. *Proc. Natural Resources Conf.,* Battelle Memorial Institute, Columbus, Ohio (1972).
22. Edmondson, W. T. and D. E. Allison. *Limnol. Oceanog., 15,* 138 (1970).
23. Crecelius, E. A. and D. Z. Piper. *Environ. Sci. Technol., 7,* 1053 (1973).
24. Johnson, R. E., A. T. Rossano, Jr. and R. O. Sylvester. *Proc. Am. Soc. Civil Engineers, 92,* 245 (1966).
25. Turekian, K. K. and M. R. Scott. *Environ. Sci. Technol., 1,* 940 (1967).
26. Bothner, M. H. and D. Z. Piper. in *Proc. of the Workshop on Mercury in the Western Environment,* R. Buhler, Ed. (Corvallis, Oregon: Oregon State University Press, 1971).
27. Held, E. E. U.S. Atomic Energy Comm. Rept. NVO-269-17 (October 1972).
28. Chow, T. J. and C. Patterson. *Earth and Planetary Sci. Letters, 1,* 397 (1966).
29. Smith, J., R. A. Nicholson, and P. J. Moore. *Nature, 232,* 393 (1971).
30. Weiss, H. V., S. Yamamoto, T. E. Crozier, and J. H. Mathewson. *Environ. Sci. Technol., 6, 7,* 644 (1972).
31. Burton, J. D. and T. M. Leatherland. *Nature, 231,* 440 (1971).
32. Bache, C. A., W. H. Gutenmann, and D. J. Lisk. *Science, 172,* 951 (1971).
33. Johnells, A. G. and T. Westermark. in *Chemical Fallout,* M. W. Miller, G. G. Berg, Eds. (Springfield: C. C. Thomas, 1969).
34. Olney, F. Unpublished manuscript.
35. Lowman, F. G., T. R. Rice, and F. A. Richards. *Radioactivity in the Marine Environment.* (Washington, D.C.: National Academy of Sciences, 1971).
36. Polikarpov, G. G. *Radioecology of Aquatic Organisms.* (New York: Reinhold, 1966).
37. Turekian, K. K. in *Impingement of Man on the Oceans,* D. W. Hood, Ed. (New York: Wiley-Interscience, 1971).

CHAPTER 4

ANALYSIS AND SPECIATION OF TRACE METALS IN WATER SUPPLIES

Patrick L. Brezonik
    Department of Environmental Engineering Sciences
    University of Florida
    Gainesville, Florida

    Water Quality Significance of Trace Metals 168
    Methods for Trace Metal Analysis 170
        Kinetic Analysis 173
        Atomic Absorption Spectrophotometry 177
        Anodic Stripping Voltammetry 180
    Trace Metal Removal by Water Treatment
        Processes 186

Although trace metal occurrences in drinking water have concerned public health officials and water treatment engineers for many years, there have been surprisingly few studies of their distribution in water supplies and of the effects of water treatment processes on trace metal levels. Further, the existing data on trace metal distributions in public water supplies has come largely from broad scale surveys and grab samples rather than in-depth studies. Much of our present ignorance on this subject can be related to the limitations of analytical techniques for trace metals. For example, most of the existing data on trace metals in U.S. water supplies [1,2] were obtained by spark emission spectroscopy, a technique useful for screening or survey purposes, but known neither for great accuracy nor for ready availability to most researchers or water treatment plant laboratories. Numerous advances have occurred in the field of trace metal analysis within the past decade, and with the recently reported statistical relationships between trace metal levels in water and food and the geography of human health problems [3], a renewed and closer study of biologically significant trace metals in drinking water and of their removal by conventional treatment processes seems overdue.

The objective of this paper is to describe recent developments in trace metal analysis that are most applicable to determination of these constituents in natural and drinking waters. Investigations of trace metal levels, speciation and removal mechanisms in treatment processes are currently under investigation in our laboratory using these analytical tools.

WATER QUALITY SIGNIFICANCE OF TRACE METALS

The presence of trace metals in water supplies has a variety of implications concerning the potability and ultimate other uses of drinking water. Table 4.1 briefly summarizes the various types of water quality implications with examples of the metals involved as well as some other trace elements. Recent renewed concern over trace metals in drinking water stems largely from human health considerations, *i.e.*, the effects of chronic low level exposure to such elements as cadmium, selenium, cobalt and beryllium. A large body of statistical evidence indicates that cardiovascular diseases are less prevalent in areas where the water supply has a high dissolved minerals content, and some investigators believe this phenomenon is related to various trace elements in water. Perhaps the greatest impetus to further studies of trace metals in water supplies has come from the recent mercury and cadmium episodes, especially the poisonings in Japan and the many reports of elevated mercury levels in waters, sediments and fish of the lower Great Lakes. While no serious episodes have been related to the occurrence of these elements in drinking water, the incidents raised the general level of public and scientific concern

Table 4.1

Classification of Trace Elements in Water Supplies
According to Water Quality Significance

Significance	Trace Element
Esthetic significance (taste and discoloration problems)	Cu, Fe, Mn, Zn
Toxic at levels found in some waters	As, B (to certain plants only), Ba, Cd, Cr, Hg, Pb, (Se?)
Toxic but present levels in water supplies are probably unimportant	Ag, Be, Bi, Ni, Sb
Probably not toxic up to ppm level (present levels: ppb or less)	Ga, Ge, Sn, Sr, Ti, V, Zr
Nutrient metals (at ppb-ppm levels), some toxic at higher levels	B, Co, Cu, Fe, Mn, Mo, Zn

about trace metals in the environment and man's exposure to them.

It should also be noted that a few toxic metals do occur relatively often in drinking water supplies at levels approaching the U.S. drinking water standards (within an order of magnitude of the standards), and several recent reports [4,5] indicate that the standards are exceeded in a small percentage of supplies. Whether the standards have always been appropriately derived and levels exceeding the standards necessarily constitute a public health hazard is another and difficult matter that we shall not consider here.

According to Table 4.1, there are eight elements of principal concern in terms of toxicity and another five which, while toxic, probably occur at insignificant levels. The physiological effects of these metals are quite varied; for example, low level cadmium exposure has been implicated in cardiovascular problems, whereas at higher levels it causes kidney and liver damage as well as the skeletal disorder osteomalacia, a severe brittling of bones. Lead affects the kidneys, nervous system, blood and brain; mercury affects the nervous system and brain, while beryllium (and selenium) have been implicated as carcinogenic. The hazard of arsenic is disputed; for many years arsenic was thought to be carcinogenic, but recent evidence disputes this [6]. It should be noted that drinking water represents only one of several sources for toxic metals; in general food is a more important source [3]. Also for specific elements and localities (*e.g.*, lead in urban areas), airborne metals may be the greatest source. The latter enter the body via the lungs, and different physiological responses may be involved for a metal depending on the mode of entry. These facts obviously compound the difficulty of setting drinking water standards for trace metals. Finally it must be pointed out that ignorance of effects of trace metals on human health probably exceeds our knowledge in this regard. This is especially true for subtle subacute effects, which by their nature are difficult to detect, let alone relate to specific causes.

If our ignorance of health implications precludes definitive conclusions regarding the "safe" levels of various trace metals in water supplies, the scattered available evidence [6,7,8] certainly provides plausible cause for concern. Thus with increased direct and indirect reuse of wastewater for drinking purposes, further information on the levels of various trace metals in water supplies and the extent to which they are removed by conventional treatment processes is clearly desirable.

METHODS FOR TRACE METAL ANALYSIS

Water supply scientists and engineers have a bewildering variety of techniques available for trace metal analyses, and selection of the most appropriate technique is not always a simple matter. Such factors as sensitivity, accuracy, ease and rapidity of analysis, potential for multi-element analysis, availability of instrumentation, cost, and the chemical form of the metal to be determined (*e.g.*, ionic as opposed to total) enter the decision calculus. Table 4.2 lists the major techniques used for analyses of trace metals in aqueous solution along with approximate levels of sensitivity. For a given method sensitivity

Table 4.2

Sensitivity of Various Methods for Metal Analyses

Method	Sensitivity	Comments
Spectrophotometry-colorimetry	10^{-5}-10^{-7} M (10^{-9} with preconcentration)	Versatile, tends to be more time-consuming, but still most useful method for certain metals
Kinetic Analysis (of reaction in which metal ion acts as a catalyst)	10^{-8}-10^{-9} M	Highly selective, sensitive; requires a small sample; not yet fully developed
Atomic Absorption Spectrophotometry (conventional) (flameless)	10^{-6}-10^{-7} M 10^{-8}-10^{-9} M	Simple, versatile; measures total metal concentration
Flame and Spark Emission Spectroscopy	10^{-5}-10^{-8} M	Gives multi-element analysis; not too accurate
Neutron Activation Analysis	10^{-9}-10^{-10} M	Requires reactor and expensive equipment
Ion Selective Electrodes	10^{-5}-10^{-6} M	Good selectivity but still too insensitive and imprecise for trace analysis
Polarography (conventional) (modified)	10^{-5}-10^{-6} M 10^{-6}-10^{-8} M	Limited to electroactive metals
Anodic Stripping Voltammetry	10^{-9}-10^{-10} M	Highly dependent on metal speciation; applicable to relatively few metals

is a function of the metal being analyzed, the matrix in which it occurs (*e.g.*, total dissolved solids directly affect sensitivity of spark emission spectroscopy and indirectly affect various other methods), and the sophistication of instrumentation (*i.e.*, sensitivity varies with brand and model).

Sensitivity is also dependent on the method of sample preparation. Some analytical procedures involve concentration of the metals from solution, and one or two orders of magnitude concentration factors can be achieved by ion exchange, distillation, solvent extraction, and coprecipitation. Of the various concentration methods, ion exchange using chelating resins [9] is probably the simplest and most desirable method for the metals and concentrations of concern in water supplies. However, in general, preconcentration is not a desirable step for it increases the possibilities of loss (incomplete recovery) and contamination [10].

Spectrophotometric methods, especially those involving solvent extraction as for example, dithizone for heavy metals, neocupreine for copper, bathophenanthroline for iron [11], are adequately sensitive for certain metals in water supplies, but these methods tend to be time-consuming and difficult to perform. Ion selective electrodes have been developed for over 20 different ions within the last 10 years, but with the possible exception of copper, the electrodes are not sufficiently sensitive to detect heavy metals at their normal levels ($<10^{-6}$ M) in water supplies.

In terms of ultimate sensitivity, it is difficult to select one method as inherently better than the others. Proponents of neutron activation analysis (NAA) have often claimed the top prize, but the method requires expensive isotope counting equipment as well as a thermal (neutron) reactor, which is available in relatively few localities. NAA requires chemical separation for certain metals, but it has the advantage of simultaneously measuring 10-20 metals in a small sample, and for many metals, sample manipulation is minimal. Bhagat, *et al.* [12] have described a procedure applicable to 13 metals at subpart per billion levels, in which activation and counting are conducted in the sealed 30-ml polyethylene bottles in which samples are collected. Thus for research purposes or for large scale monitoring studies NAA may be useful or even the method of choice, but it would not seem applicable to routine metal analyses in water treatment plants.

The claim of greatest sensitivity made by proponents of neutron activation analysis has been challenged recently by several analytical developments that are potentially more widely available for routine analysis. These methods include atomic absorption spectrophotometry, especially in its flameless modifications, kinetic or reaction rate analysis, in which the metal acts as catalyst of a chemical (colorimetric) reaction, and anodic stripping voltammetry. Each of these methods has a sensitivity of at least $10^{-8}-10^{-9}$ M for some metals. Because of their great potential for ultra trace determinations, these techniques will be discussed later in further detail.

It should also be noted that 10^{-9} M sensitivity is not always required and other less sensitive methods may be more appropriate under these circumstances. Table 4.3 summarizes the available data for trace metals in water supplies and drinking waters, as reported by Durfor and Becker [1], Kroner and Kopp [2], Kopp [4], Taylor [13], Barnett, et al. [14] and others. More detailed discussions of trace metal levels in drinking water have been presented in the above papers and elsewhere. Of interest here is the point that concentration ranges vary widely from metal to metal; some metals occur in the subpart per billion (ppb) range, others in the ppb range and still others in the sub-ppm range. Quite obviously, then, no single analytical method is best for all metals, and the list of usual and potential analytical methods for the metals listed in Table 4.3 reflects this fact.

Kinetic Analysis

Conventional colorimetric analyses, in which the metal ion combines stoichiometrically with some (usually organic) reagents to form a colored species, are usually limited to the sub-ppm range. For a few metals, such as copper and iron, color products with especially high absorptivities may allow direct analysis to ppb levels, but without some sort of concentration step this is unusual. However, the sensitivity of colorimetric analysis can be extended several orders of magnitude when the sought for species acts as a catalyst for some reaction whose reactants are present in higher concentrations rather than as a stoichiometric reactant to form some colored species directly. The (trace metal) catalyst then

Table 4.3

Trace Metal Levels in U.S. Water Supplies
and Usual Procedures for their Analysis

Metal	Range[a]	Usual Method[b]	Potential Methods
Antimony	1-100	ES	AA
Arsenic	0-100	C	ASV
Barium	0.7-900	ES	
Beryllium	0.01-0.7	ES	
Bismuth	0.04-70	ES	ASV
Boron	0-1000	C	
Cadmium	0.4-60	ES,AA,ASV	
Chromium	0.3-40	C,AA,ES	
Cobalt	0.3-30	C,ES	ASV
Copper	0-600	C,AA,ASV	IE
Germanium	?	C,ES	
Iron	1->300	C,AA	K
Lead	1-400	C,ES,AA,ASV	IE
Manganese	0->50	C,AA	K
Mercury	0.01-30	FAA	
Molybdenum	0.4-9200	C,ES,AA	K
Nickel	0.4-400	ES,C	AA
Selenium	0-10	C	K
Silver	0-2	ES	K
Tin	0.3-30	ES,C	AA
Titanium	4-50	ES	AA
Vanadium	0.4-70	C,ES	AA
Zinc	60-7000	C,AA,ASV	IE
Zirconium	?	ES	C

[a]Estimated from U.S.P.H.S. water quality network data [1,2,13]. Values in ppb.
[b]ES = emission spectroscopy, C = colorimetry, AA = atomic absorption, ASV = anodic stripping voltammetry, FAA = flameless atomic absorption, IE = ion selective electrode, K = kinetic analysis.

produces a (color) change far out of proportion to its concentration. The rate of reaction, which can be measured by recording color production or loss against time or by determining the amount of color formed (or remaining) after a specified reaction time, is directly proportional to catalyst concentration if other conditions are maintained constant. This is the basis of kinetic or reaction rate analysis, a rapidly expanding area of analytical chemistry [15,16,17].

The advantages of kinetic analysis include: (1) simple rapid analysis (in many cases), (2) small sample volumes required for catalytic reactions, (3) great sensitivity, enabling preconcentration steps to be eliminated, and (4) great selectivity of many catalysts for specific reactions, thus minimizing the need for separations or possibility of interferences. Application of reaction rate methods to trace analyses of natural water constituents has been hindered in the past by instrumentation inadequacies, but several manufacturers are now marketing instruments specifically designed for reaction rate techniques and programmed for easy use as well as direct concentration readout [18]. Many reaction rate methods require instrumentation no more sophisticated than a good spectrophotometer, and some methods can be adapted to use AutoAnalyzer equipment. The growing number of applications of reaction rate methods to trace metal analyses have been reviewed by Rechnitz [19], Guilbault [20] and Greinke and Mark [21].

Figure 4.1 illustrates an AutoAnalyzer adaptation of a reaction rate method for molybdenum, a technique showing promise of applicability to natural water measurements. The manifold was developed by Carriker [22] from the manual method of Wilson [23]. The basic reaction involved is oxidation of iodide to iodine by peroxide or perborate (in this case); molybdenum acts as a catalyst at ppb levels (Figure 4.2) and the amount of iodine (measured at 350 nm) or blue starch-iodine complex (measured at 550 nm) formed in a given reaction time is proportional to molybdenum concentration. With further modification and care, the method could probably be made sensitive to Mo levels as low as 0.1 ppb without preconcentration or sample preparation.

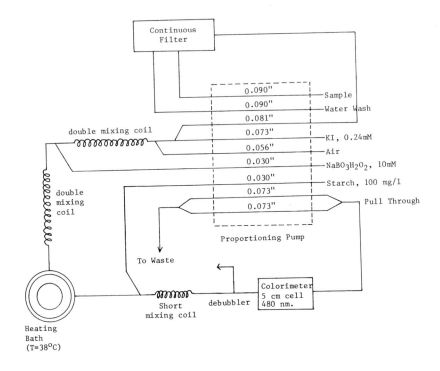

Figure 4.1. AutoAnalyzer flow diagram for molybdenum by perborate-iodate reaction rate method.

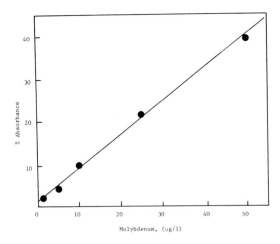

Figure 4.2. Standard curve for molybdenum using AutoAnalyzer method described in Figure 4.1.

One cautionary note should be mentioned with regard to reaction rate methods. Although catalysis is usually quite selective, some reactions may be catalyzed by metals other than the sought for species. In other cases, catalyst specificity is less a problem than other direct reaction interferences. For example, in the molybdenum procedure described above, interference arises from the presence of oxidants such as chlorine, which will form iodine even in the absence of molybdenum. Thus application of this method to treated drinking water would require pretreatment to remove chlorine.

Atomic Absorption Spectrophotometry

When atomic absorption (AA) spectrophotometry was first introduced commercially about 10 years ago, it was widely acclaimed, especially by instrument manufacturers, as "virtually interference-free." There have been many advances in AA techniques in the first decade of its use, and atomic absorption has doubtless become the most popular method for major and trace metal analysis in fields ranging from soil and water chemistry to metallurgy. Recent developments of nonflame atomizers like the graphite furnace have increased sensitivity by several orders of magnitude, and direct analysis of water samples for a wide range of trace metals is now possible.

Along with the widely discussed advances and the widespread adoption of AAS, a growing literature on problems and interferences has also arisen. Atomic absorption is still regarded by experts as relatively interference-free, and *with care* it will yield accurate results (often better than other methods). A great volume of literature exists on the theory and application of atomic absorption techniques (*e.g.*, references 24-27). On the other hand it is probably still a common impression among workers casually acquainted with AA that the technique is nearly foolproof and interference-free. Thus it is important to consider the interferences that do exist, especially in light of the widespread use of AAS for routine analysis.

The mechanisms of interference can be grouped into spectral, chemical and matrix effects as summarized in Table 4.4, and

Table 4.4

Summary of Interference Mechanisms
in Atomic Absorption Spectrophotometry

	Interference
Spectral	Molecular absorption Line (element) absorption
Chemical	Oxide formation Refractory compounds formed with phosphate, silicates Intermetallic compounds Ionization suppression or enhancement by alkali metals
Matrix	Viscosity, acid content, surface tension—all affect efficiency of atomization Light scattering caused by high solids

specific examples of interferences relevant to water analysis are detailed in Table 4.5. Although a few line absorption interferences have been found [28] none of these would seem important for water analysis. Molecular absorption is now known to be more common, especially in the ultraviolet region, but this can usually be overcome by using a hotter flame. Chemical interferences are probably the best-studied problem in AAS, and methods of avoiding them (*e.g.*, use of a hotter nitrous oxide flame, addition of a chemical suppressor) are widely described and accepted. Matrix effects have received less attention; of these light scattering by high solids would seem the most serious problem for trace analysis. This problem has been recognized as especially serious in heated graphite atomizers, and manufacturers recommend use of a deuterium continuum source for automatic background correction in these circumstances. But the seriousness of this problem in flame AAS analysis of trace metals with resonance lines in the far UV is unknown and bears further study.

Table 4.5
Interferences for Some Metals in Atomic Absorption Spectrophotometry

Elements	Problem	Solution	References
Al,Ba,Ca,Cr,Mn, Mo,Pb,Sn,Sr,Ti,V	In samples with variable amounts of alkali metals there is ionization of these metals in the flame; especially serious in N_2O flame	Add easily ionized Cs or K in excess (100-1000 mg/l) to samples and standards as ionization suppressor	28
Al,B,Be,Ge,Mo, Si,Ti,V,W,Zr	Metal oxide formation in air–C_2H_2 flame	N_2O-C_2H_2 flame	29,30
Ba	CaO (molecular absorption)	N_2O flame (decomposes CaO)	31
Ca,Fe,K,Mg, Mn,Zn	High sodium decreases efficiency of atomization	Dilute samples or add same amount of Na to all samples and standards	32
Ca,Mg	Phosphate and silicates form refractory compounds in air–C_2H_2 flame	Sr or La–HCl added to samples and standards	33
Cr	Ni in air–C_2H_2 flame	Use fuel-rich flame or N_2O flame	34
Fe,Mn	Silica	Add 200 ppm Ca	35
Mg	Aluminum (intermetallic) compound formation in air–C_2H_2 flame	N_2O flame	36
Mo		Add NH_4Cl or use N_2O flame	37
Si	Vanadium spectral line interference at 250.6 nm	Measure at 251.6	38
Elements with analytical resonance line in far UV (i.e., ~<250nm)	Scattering by high solids (particles) in flame; especially serious near detection limits of elements	Filter out solids; higher flame temp.; deuterium or H_2 continuum source for background correction	25,39,40

Anodic Stripping Voltammetry

Although anodic stripping voltammetry (ASV) is applicable to a relatively small group of metals, it has recently received considerable attention from analytical water chemists because of its great sensitivity and the possibility of its use in determining metal ion speciation [41]. The theory involved in ASV has been thoroughly described [42,43]. Briefly, the method is a two-step process of concentration and analysis. The first step involves reduction of metal ions to their elemental state and their simultaneous deposition onto a mercury drop or film electrode which is maintained at a low potential for a controlled period of time under constant stirring conditions. This is termed the pre-electrolysis, cathodic or deposition step, and in effect the ions are concentrated from solution into a mercury amalgam. The second or analytical step involves application of a linearly increasing potential and measurement of the resulting current-potential (i *vs* v) curve. This is called the anodic or stripping step, and current peaks appear at potentials where the plated metals are oxidized. Peak current is proportional to original (solution) metals concentration (assuming constant deposition and stripping conditions), and peak potential is indicative of the metal involved.

Although a few nonamalgam forming metals such as cobalt [Brauner, unpublished data] have been measured by ASV, the method is generally limited to those metals that do form amalgams. Those most relevant to natural water studies are cadmium, copper, lead and zinc. Mercury could also be determined by ASV with a graphite electrode, but other methods (*e.g.*, flameless atomic absorption) are probably better suited for trace level mercury analysis. For the few metals amenable to analysis by ASV the method does offer some significant advantages: (a) high sensitivity (10^{-9} M or $\sim$ 0.1 ppb or better), (b) large dynamic range obtainable by varying deposition time, (c) relatively simple and inexpensive instrumentation (approximately $2000-5000 excluding recorder), and (d) suitability for field work (portability).

Proponents of ASV also suggest that the method can be used to distinguish among possible metal species in solution since some forms (*e.g.*, certain organic complexes) are not reducible at the mercury drop whereas other forms are. By changing solution conditions and measuring changes in peak potential (E_p) and peak current (i_p), one should (theoretically) be able to make inferences about metal speciation—in a manner similar to the way one obtains information on complex formation from conventional polarography, except at much lower concentrations. For example, the complexation of metals with weak organic acids such as EDTA and amino acids is pH dependent, and the strength of complexation increases with pH. Since most known organic complex formers in natural waters are weak acid salts, the inference is generally made that metal-organic interactions occur *in situ* under neutral or alkaline conditions but that under acidic pH (pH<~3) conditions the metals exist as the free hydrated species. Most metal-organic complexes are thought to be nonreducible at the mercury electrode and hence not measurable by ASV. The difference in peak current (i_p) under ambient (neutral) conditions and under acidic (pH<3) conditions has then been interpreted [41,44,45] as representative of the amount of metal ion complexed by organic species or more generally as representative of the metal present in "nonlabile" complexes of some undetermined nature [41,44].

There are at least two circumstances under which the above theory would be invalid: (1) the presence of an organic substance that forms metal complexes more strongly under acidic conditions than at neutral or alkaline pH or forms complexes independent of pH (at least within a broad pH range), and (2) the presence in solution of surface active substances that coat the mercury electrode and prevent metal ion deposition or metal oxidation. If sorption effects are pH dependent, which seems likely for surface active substances occurring in natural waters, interpretation of ASV results in terms of metal ion speciation will be in error.

A wide variety of soluble and colloidal materials with surface active characteristics occur in natural waters, among which are detergents, proteins, oils and fats. For example, gelatin may be considered representative of colloidal proteinaceous matter present in natural waters as a result of organismic activities. Figure 4.3 illustrates the effects of gelatin on cadmium analysis by ASV under neutral and acidic conditions. No effects were

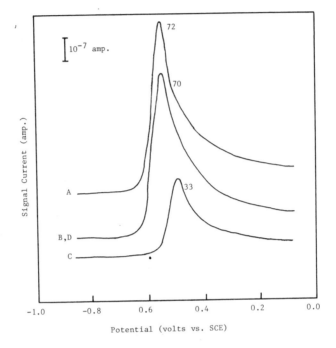

Figure 4.3. Effect of gelatin on cadmium analysis by ASV. **A.** control, 2×10^{-6} M Cd; **B.** control plus 20 mg/l gelatin; **C.** solution B acidified to pH 3 with HNO_3; **D.** same solution after addition of NaOH to return pH to 7.

observed under neutral conditions at up to 20 mg/l of gelatin, but when the pH was decreased to 3.0, cadmium peak current was depressed by more than 50%. The effect was instantly reversible upon raising the pH back to neutrality. It is not possible to conclude from this figure alone whether the phenomenon resulted from increased complexation of cadmium by gelatin under acidic conditions or from increased sorption of gelatin onto the mercury drop electrode as pH is decreased. However, gelatin is a well-known maximum suppressor in polarography and is known to sorb onto mercury. Regardless of the mechanism, it is clear that the co-occurrence of gelatin-like substances and chelating acid salts such as amino acids will obviate simple interpretation of changes in i_p vs. pH as indicative of the relative amounts of free and complexed metal existing under *in situ* conditions.

Figure 4.4 shows that the effects of interferences on ASV results depend also on the metal involved. When the soluble enzyme alkaline phosphatase is added to a solution containing both cadmium and copper at pH 7, i_p for copper is depressed to about 60% of the control at enzyme concentrations as low as 2 mg/l, but cadmium is affected only slightly up to an enzyme level of 36 mg/l. However, when the pH is decreased to ~3.0, the effect on the metals is essentially reversed. A simple interpretation of these results is not readily apparent, but again the

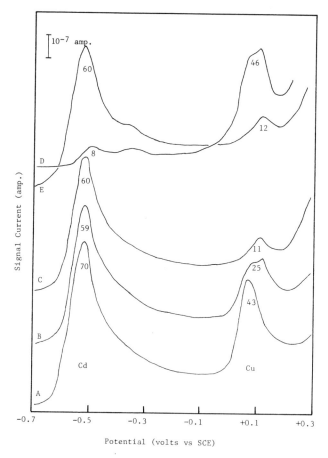

Figure 4.4. Effect of alkaline phosphatase (APase) on cadmium and copper measurement by ASV. Cu and Cd = 1×10^{-6} M; ionic strength = 0.1; plating time = 3.0 min. **A.** no APase (control), **B.** APase = 2 mg/l, **C.** APase = 36 mg/l, **D.** pH decreased from 7.0 in A-C to 3.05, **E.** pH raised back to 6.0. Numbers under peaks are i_p (peak current) in 10^{-8} amp.

results place an obvious cloud over the application of ASV to *in situ* speciation studies, especially in polluted (organic rich) waters.

In controlled laboratory studies where influences of unknown trace organics can be minimized ASV may be a more useful technique to study metal ion speciation and reactivity. Figure 4.5 illustrates peak current against pH for a 10^{-6} M Cd(II) solution in 0.1 M KNO_3. Superficial observations might suggest that the decrease in i_p reflects decline in Cd(II) solubility as K_{sp} for cadmium hydroxide is exceeded. Since no visible precipitate occurred in the electrolysis vessel during this experiment, the curve might also be interpreted to indicate that Cd^{2+} and $CdOH^+$ ions are reducible species, whereas at the mercury electrode soluble species $Cd(OH)_2°$ is not. Qualitatively these

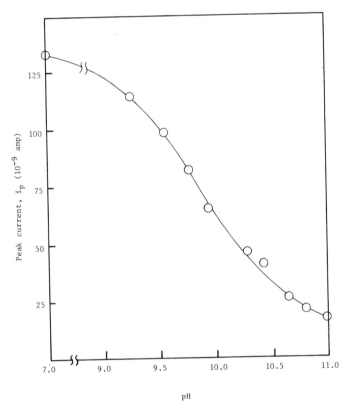

Figure 4.5. Peak current (i_p) *vs.* pH for 1×10^{-6} M Cd in 0.1 M KNO_3

conclusions may be correct, but comparison of the experimental curve with the theoretical distribution of Cd species as a function of pH (*e.g.*, Figure 4.6) reveals serious quantitative discrepancies.

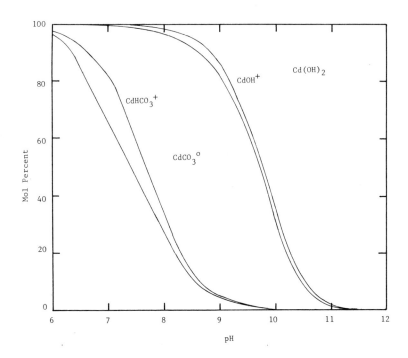

Figure 4.6. Distribution diagram for cadmium species as function of pH in water with total carbonate = 2×10^{-3} M.

It is obvious that the rate of decrease in i_p with pH is far lower than the rate of change in any Cd species or combination of species.

The possible application of ASV to determination of metal ligand stability constants and ultimately to metal ion speciation remains an interesting possibility [46,47]. However, many questions remain about the relationships among metal ion speciation, solution conditions, metal reducibility, and reversible as opposed to irreversible character of metal redox reactions and their effects on the diagnostic parameters, E_p and i_p, used to evaluate ASV results.

The above somewhat pessimistic note should not be interpreted to mean that the author regards anodic stripping voltammetry as a useless method for trace metal determinations. For the few metals to which it is applicable, ASV is highly sensitive, and if organic interferences are absent (or eliminated by digestion) the method should be useful for total metal determinations. Indeed since drinking waters tend to be low in dissolved organic contaminants ASV may be particularly useful to monitor toxic metals like cadmium and lead in water supplies. Likewise, the results of Brauner et al.[48] described in part above do not completely eliminate the application of ASV to speciation studies. Rather they do suggest that caution be used in interpreting *in situ* ASV determinations in terms of metal-ligand interactions and that the effects of sorption and solution conditions on mercury electrode surfaces are a potentially serious problem in ASV applications.

TRACE METAL REMOVAL BY WATER TREATMENT PROCESSES

Relatively few studies have been conducted on the degree of trace metal removal achieved by water treatment processes. Much of the available information has been obtained from grab samples of raw and treated water from municipal water treatment plants [e.g., references 1,14]. The results presented by Andelman [49] indicate the dangers involved in this approach; water supplies, especially surface supplies, can hardly be regarded as time invariant sources. However, a few general conclusions are apparent from these surveys. For example, removal efficiency is highly variable among the various metals, and is definitely correlated with metal solubility characteristics. Thus monovalent ions such as cesium and unionized species such as boron (probably present as boric acid, H_3BO_3) are poorly removed, whereas relatively insoluble divalent ions such as copper and strontium are generally much lower in treated than raw water. Certain metals consistently are higher in treated water than in the raw supply. Of these aluminum is most notable and obviously reflects its use as alum in coagulation. Copper, zinc, cadmium and iron may be higher in distribution systems because of corrosion than in the raw water supply itself.

Finally many metals show variable responses, with high removal efficiencies in some plants and little or no removal in others. At least three explanations could account for this behavior: (1) sampling inadequacies for temporally varying trace metals leading to erroneous conclusions regarding percentage removals, (2) differential removal of metals by various treatment processes, and (3) differential removal caused by varying solution conditions in various water supplies. Probably all three factors are important in evaluating trace metal removal data from the literature. The potential health implications of trace metals in drinking water coupled with the meager information on trace metal removal by conventional treatment practices available in the literature suggest that detailed laboratory and pilot plant studies of this subject, including such aspects as removal mechanisms and relationships between metal speciation and removal efficiency, are long overdue.

Certainly softening plants would tend to have different trace metal removal efficiencies than would coagulation-filtration or simple filtration plants. On the other hand, the form in which a trace metal exists in a water supply will also affect removal efficiency by softening and coagulation. For example, metals that form insoluble hydroxides and/or carbonates (*e.g.*, cadmium, copper, lead, strontium, zinc) and are present in the raw water as inorganic ions should be removed with high efficiency by lime-soda softening since the pH will be highly alkaline during softening and the metal carbonates or hydroxides should coprecipitate with calcium carbonate and magnesium hydroxide sludge. But if these metals were present in the water supply complexed to organic coloring matter, softening would be less efficient at their removal since color removal is best achieved by coagulation with alum or iron salts under acidic conditions.

Solution pH will also determine the extent to which trace metals present as simple hydrated ions or inorganic complexes will be removed by alum coagulation for turbidity removal. As indicated by Stumm and Morgan [50], hydrolysis products of multivalent ions are adsorbed more readily at solution-particle interfaces than nonhydrolyzed metal ions. Quantitatively this can be explained by a variety of considerations, such as hydrolyzed species are larger and less hydrated, and replacement of an aquo group by a hydroxo group in the metal atom's coordination sphere reduces the effective charge of the ion and renders the complex more hydrophobic. Figure 4.7 shows a simple case of this phenomenon in the sorption of cadmium by kaolinite. Sorption was not measurable under acidic conditions but occurred rapidly when the pH was raised to 9.5. It should be noted that

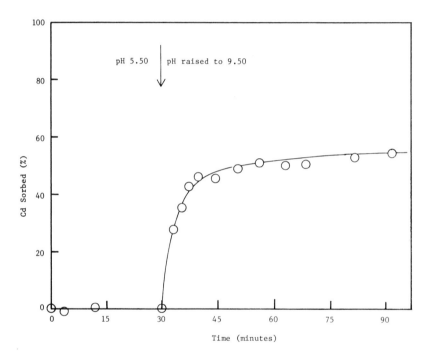

Figure 4.7. Kinetics of cadmium sorption onto kaolinite at pH 5.50 and 9.50. 9.50. Initial cadmium = 1×10^{-6} M; kaolinite = 10 mg/l.

pK_1 for $CdOH^+$ is approximately 9.0, indicating that $CdOH^+$ and further hydrolysis products predominate over Cd^{2+} at pH 9.5. This phenomenon has important implications for trace metal removal in treatment processes and suggests that metals will be readily removed by solid surfaces (alum floc, softening sludges, sand filters) provided the pH is high enough to cause metal hydrolysis. Thus metal ion speciation is an important consideration in removal efficiencies of treatment processes and analytical developments that allow determination of speciation should be pursued.

ACKNOWLEDGMENTS

The experimental part of this paper is based on work conducted at the Swiss Federal Institute for Water Resources and Water Pollution Control under an NSF Science Faculty Fellowship. The cooperation of W. Stumm, Institute Director, is gratefully acknowledged. The ASV studies were done in cooperation with P. A. Brauner who was on sabbatical from Simmons College. Her contribution to this paper was invaluable but typically she declined authorship.

REFERENCES

1. Durfor, C.N. and E. Becker. "Public Water Supplies of the 100 Largest Cities in the United States, 1962." Water Supply Paper 1812, U.S. Geol. Surv., Washington, D.C. (1964).
2. Kroner, R. C. and J. F. Kopp. *J. Amer. Water Works Assoc., 57,* 150 (1965).
3. Maugh, T. H. *Science, 181,* 253 (1973).
4. Kopp, J. F. *Proc. Conf. Trace Substances in Environ. Health, 3,* 59 (1969).
5. McCabe, L. J., J. M. Symons, R. D. Lee and G. G. Robeck. *J. Amer. Water Works Assoc., 62,* 670 (1970).
6. Panel on Public Water Supplies. Draft report to National Academy of Science, National Academy of Eng. (1972)(mimeo).
7. Federal Water Pollut. Control Admin. "Water Quality Criteria," Nat. Tech. Advis. Comm. Report, U.S. Govt. Printing Office, Washington, D.C. (1968).
8. McKee, J. and H. W. Wolf. "Water Quality Criteria," 2nd edition, Publ. 3-A, State Water Quality Control Board, Sacramento, Calif. (1963).
9. Fisher, R. P. "An Investigation of Atomic Absorption Analysis of Mill Effluent Metal Ion Content," (New York: National Council of the Pulp and Paper Industry for Air and Stream Improvement, Tech. Bull. 262, 1972).
10. Robertson, D. E. *Anal. Chem., 40,* 1067 (1968).
11. American Public Health Association. *Standard Methods for the Examination of Water and Wastewater.* (New York: American Public Health Association, 1971).
12. Bhagat, S. K., W. H. Funk, R. H. Filby and K. R. Shah. *J. Amer. Water Works Assoc., 61,* 61 (1969).
13. Taylor, F. B. *J. Amer. Water Works Assoc., 55,* 619 (1963).

14. Barnett, P. R., M. W. Skougstad and K. J. Miller. *J. Amer. Water Works Assoc., 61*, 61 (1969).
15. Brezonik, P. L. in *Water and Water Pollution Handbook*, Vol. 3, L. L. Ciaccio, Ed. (New York: Marcel Dekker, 1972), p. 831.
16. Malmstadt, H. V., E. A. Cordos and C. J. Delaney. *Anal. Chem., 44*, 26A (1972).
17. Mark, H. B., Jr. and G. Rechnitz. *Kinetics in Analytical Chemistry.* (New York: Interscience, 1968).
18. Malmstadt, H. V., C. J. Delaney and E. A. Cordos. *Anal. Chem., 44*, 79A (1972).
19. Rechnitz, G. A. *Anal. Chem., 40*, 455R (1968).
20. Guilbault, G. *Anal. Chem., 42*, 334R (1970).
21. Greinke, R. A. and H. B. Mark, Jr. *Anal. Chem., 44*, 295R (1972).
22. Carriker, N. Unpublished report, University of Florida, Gainesville (1973).
23. Wilson, A. M. *Anal. Chem., 38*, 1784 (1966).
24. Christian, G. D. and F. J. Feldman. *Atomic Absorption Spectroscopy. Applications in Agriculture, Biology and Medicine.* (New York: Wiley, 1970).
25. Kahn, H. in *Trace Inorganics in Water.* (Washington, D.C.: American Chemical Society, Advanced Series in Chemistry 73) 1968), p. 183.
26. Robinson, J. W. *Atomic Absorption Spectroscopy.* (New York: Marcel Dekker, 1966).
27. Winefordner, J. D. and T. J. Vickers. *Anal. Chem., 44*, 150R (1972).
28. Pickett, E. E. and S. R. Koirtyohann. *Anal. Chem., 41*, 28A (1969).
29. Manning, D. C. *Atomic Absorp. Newsletter, 5*, 127 (1966).
30. Manning, D. C. *Atomic Absorp. Newsletter, 6*, 35 (1967).
31. Capacho-Delgado, L. and S. Sprague. *Atomic Absorp. Newsletter, 4*, 363 (1965).
32. Ramirez-Muñoz, J. *Anal. Chem., 42*, 517 (1970).
33. Fishman, M. *Atomic Absorp. Newsletter, 5*, 102 (1966).
34. Dyck, R. *Atomic Absorp. Newsletter, 4*, 170 (1965).
35. Platte, J. A. and V. M. Marcy. *Atomic. Absorp. Newsletter, 4*, 289 (1965).
36. Menzies, A. C. *Anal. Chem., 33*, 1226 (1961).
37. Mostyn, R. A. and A. F. Cunningham. *Anal. Chem., 38*, 121 (1966).
38. Fassel, V. A., J. D. Rasmusson, and T. G. Cowley. *Spectrochim. Acta, 23B*, 579 (1968).
39. Billings, G. K. *Atomic Absorp. Newsletter, 4*, 357 (1965).
40. Koirtyohann, S. R. and E. E. Pickett. *Anal. Chem., 38*, 1087 (1966).
41. Mancy, K. H. *Proceedings International Water Pollution Research Conference*, Vol. 6 (New York: Pergamon Press, 1973).
42. Barendrecht, E. in *Electroanalytical Chemistry*, Vol. 2, A. V. Bard, Ed. (New York: Marcel Dekker, 1967).
43. Shain, I. in *Treatise on Analytical Chemistry*, Part I. I. M. Kolthoff, Ed. (New York: Interscience, 1964).
44. Allen, H. E., W. R. Matson and K. H. Mancy. *J. Water Pollut. Control Fed., 42*, 573 (1970).

45. Siegel, A. in *Organic Compounds in Aquatic Environments,* S. D. Faust and J. V. Hunter, Eds. (New York: Marcel Dekker, 1971), p. 265.
46. Gübeli, A. O. and J.-P. Retel. *Helv. Chim. Acta, 55,* 1429 (1972).
47. Stumm, W. and H. Bilinski. *Proceedings International Water Pollution Research Conference,* Vol. 6 (New York: Pergamon Press, 1973).
48. Brauner, P. A., P. L. Brezonik and W. Stumm. "Effects of Adsorption on Trace Metal Analysis by Anodic Stripping Voltammetry," manuscript in preparation.
49. Andelman, J. B. in *Chemistry of Water Supply, Treatment and Distribution,* A. J. Rubin, Ed. (Ann Arbor, Mich.: Ann Arbor Science Publishers, 1973).
50. Stumm, W. and J. J. Morgan. *Aquatic Chemistry.* (New York: Wiley-Interscience, 1970).

CHAPTER 5

OXYGENATION OF FERROUS IRON(II) IN HIGHLY BUFFERED WATERS

Mriganka M. Ghosh
    Department of Civil Engineering
    University of Maine
    Orono, Maine

    Background 194
        Oxidation Kinetics of Iron(II) 194
        Solubility of Iron(II) 196
        Alkalinity and pH Control 197
        Buffer Capacity 200
        Calculations for Supersaturation 201
    Experimental Method 204
    Results and Discussion 206
        Buffer Capacity Effects 206
        Supersaturation Effects 212
        Application of CSTR Theory 212
    Conclusions 215

The design of iron-removal facilities traditionally has been based on empirical relationships. The inadequacies of existing design procedures are well documented [1] by numerous case histories where such facilities failed to meet the overall treatment objectives. The design of chemical reactors for the removal of soluble iron in natural waters is quite complex. However, kinetic data developed in simulated iron-removal systems can be used to develop rational design criteria for full scale plants.

Iron occurs naturally in two oxidation states, namely, the reduced or ferrous form, Fe(II), and the oxidized or ferric form,

194 Aqueous-Environmental Chemistry of Metals

Fe(III). Both soluble iron species and insoluble hydrolyzed iron may be present in aquatic systems. Iron(II) is thermodynamically unstable in the presence of oxygen. Therefore, in natural surface waters it is found mostly as hydrated ferric oxides having low solubilities.

The most common method for the removal of iron from groundwater consists of aeration of the water followed by sedimentation and filtration. Aeration raises the pH of the water by stripping CO_2 and introduces oxygen for the oxidation of Fe(II) to Fe(III). The precipitate of hydrolyzed Fe(III) is then removed by sedimentation and filtration.

Decomposition of organic matter by soil bacteria can often create low pH and anoxic conditions in subsurface waters. This may result in the dissolution and chemical reduction of Fe(III) from minerals to the soluble Fe(II) form. Another important aspect of the chemistry of iron is the formation of complex ions with inorganic as well as organic ligands. The difficulty encountered in removing iron from certain natural waters is often associated with this phenomenon.

The purpose of this study was to investigate the kinetics of iron(II) oxygenation in highly buffered waters over the pH range of 5.5 to 7.2. The oxygenation rate constants were determined on synthetic waters using batch systems. Continuous flow studies were run to evaluate the applicability of continuous stirred-treatment-reactor (CSTR) theory in predicting the removal of Fe(II) from natural waters.

BACKGROUND

Oxidation Kinetics of Iron(II)

The iron oxidation process using dissolved oxygen has been studied rather extensively by different investigators. The stoichiometric relationship describing the oxidation of iron(II) to the ferric form and its subsequent hydrolysis in natural waters is

$$2 \text{ Fe}^{2+} + 5 \text{ H}_2\text{O} + \tfrac{1}{2} \text{ O}_2 \rightleftharpoons \text{Fe(OH)}_3 \text{(s)} + 4 \text{ H}^+. \tag{1}$$

It has been shown that in the neutral pH range the rate of iron oxygenation is first-order with respect to the concentration of both iron(II) and oxygen, and second order with respect to hydroxide ion concentration. The kinetics proposed by various investigators are shown in Table 5.1. Based on experimental work obtained for acidic systems, Singer and Stumm [5] concluded that oxygenation rate is independent of pH below 3.0 and that there is a significant departure from the second order dependence of the rate on pH in the range 3.0 to 4.5.

Table 5.1

Kinetics of Iron(II) Oxidation

Equation	Reference
$-\dfrac{d\,[Fe(II)]}{dt} = \dfrac{k\,[Fe(II)]\,[O_2]}{[H^+]^2}$	2
$\dfrac{d[Fe^{3+}]}{dt} = k\,[Fe^{2+}]\,[O_2]$ (k dependent on pH)	3
$-\dfrac{d\,[Fe(II)]}{dt} = k\,[Fe(II)]\,[O_2]\,[OH^-]^2$	4

A number of investigators have observed that alkalinity has an effect on the oxygenation rate of Fe(II). Applebaum [1] reported that high alkalinity and high solids generally favored oxygenation. Longley [6] made similar observations with natural groundwaters. Ghosh, et al. [7] performed studies on the rate of oxidation and removal of Fe(II) in natural groundwaters in Illinois. They observed a first-order dependence of the rate on pH. However, a definite correlation was observed between the rate of oxidation and alkalinity. Stumm and Lee [4] also observed that the oxygenation of Fe(II) was slower in solutions of lower alkalinity. This was attributed to the slow response of the

HCO_3^--CO_2 buffer system to localized acidity changes induced by the oxygenation reaction. The rates observed by Ghosh, *et al.* [7] in groundwaters were approximately one order of magnitude less than predicted by the kinetics proposed by Stumm and Lee [4]. One possible explanation is the lower level of alkalinities in natural waters as compared with the synthetic waters used by Stumm and Lee. In discussing the field data reported by Ghosh, *et al.*, Stumm and Singer [8] proposed that the presence of inhibitors in natural groundwaters, such as organic substances, could very well retard the oxidation process in accordance with the reaction sequence

$$Fe(II) + ¼ O_2 + \text{organic} = Fe(III) \text{ - organic complex} \quad (2)$$

$$Fe(III) \text{ - organic complex} = Fe(II) + \text{oxidized organic} \quad (3)$$

$$Fe(II) + ¼ O_2 + \text{organic} = Fe(III) \text{ - organic complex} \quad (4)$$

In such cases, the Fe(II)-Fe(III) redox couple acts as an electron-transfer catalyst for the oxidation of organic matter. Depending on the relative rates of Fe(II) and Fe(III) reduction by organics, a significant retardation of the overall oxygenation of Fe(II) is possible. Oldham and Gloyna [9] suggested a similar scheme for iron transport in soils in the presence of humic substances.

Solubility of Iron(II)

Iron exists principally in the reduced form in most natural groundwaters and in the hypolimnetic waters of lakes and reservoirs. In the absence of carbonate and sulfide species the solubility of Fe(II) is controlled by $Fe(OH)_2(s)$. However, in natural waters having considerable alkalinities and within the pH range of 7 to 10, the solubility is governed primarily by $FeCO_3(s)$. Figure 5.1 shows the solubility of Fe(II) in the absence of carbon dioxide and sulfide species. The equilibrium data used in constructing the solubility diagram is shown in Table 5.2. The effect of 5×10^{-3} M total carbonate species, C_T, on Fe(II) solubility has also been superimposed on the same figure. If the groundwater exhibits a high degree of supersaturation with respect to $FeCO_3(s)$,

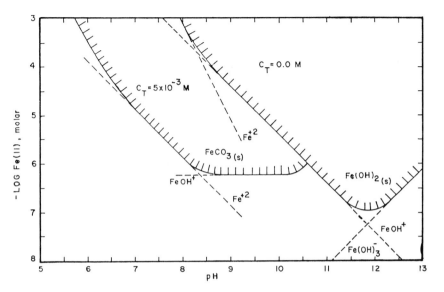

Figure 5.1. Solubility of iron(II).

the removal of Fe(II) by oxygenation may be caused both by the conversion of Fe(II) to $Fe(OH)_3(s)$ and by precipitation as $FeCO_3(s)$:

$$Fe^{2+} + HCO_3^- = FeCO_3(s) + H^+. \tag{5}$$

In waters having high alkalinities, this dual mechanism of iron removal is a distinct possibility.

Alkalinity and pH Control

The control of pH and alkalinity was of utmost importance in this study. During the course of iron oxidation the hydrolysis of the oxidized species of iron decreases the pH. In order to maintain constant pH and constant alkalinity it was necessary to vary the partial pressure of CO_2 in the gas mixture. The methods used for maintaining constant pH and alkalinity and constant pH and total carbonate (C_T) are outlined below. It should be realized that the formation of large quantities of complexes (carbonato

Table 5.2
Equilibria for Solubility of Iron(II)

Equation Number	Reaction	Equilibrium Constant at 25°C	pK	Reference
1	$Fe(OH)_2(s) = Fe^{2+} + 2\ OH^-$	8×10^{-16}	15.1	10
2	$Fe(OH)_2(s) = FeOH^+ + OH^-$	4×10^{-10}	9.4	10
3	$Fe(OH)_2(s) + OH^- = Fe(OH)_3^-$	8.3×10^{-6}	5.1	11
4	$FeCO_3(s) = Fe^{2+} + CO_3^=$	2.1×10^{-11}	10.68	12
5	$FeCO_3(s) + OH^- = FeOH^+ + CO_3^=$	1×10^{-5}	5	12
6	$FeCO_3(s) + H^+ = Fe^{2+} + HCO_3^-$	4.48×10^{-1}	0.35	Computed
7	$CO_2(g) + H_2O = H_2CO_3^*$	3.38×10^{-2}	1.47^a	13
8	$H_2CO_3^* = H^+ + HCO_3^-$	4.45×10^{-7}	6.35	14
9	$HCO_3^- = H^+ + CO_3^=$	4.69×10^{-11}	10.33	15

[a]Henry's law constant: $K = [H_2CO_3^*]/p_{CO_2}$, (M/atm)

$H_2CO_3^*$ represents the sum of dissolved $[CO_2]$ and $[H_2CO_3]$.
But $[H_2CO_3] \ll [CO_2]$, and therefore $[CO_2] = [H_2CO_3]$.

and hydroxo) may affect the equilibrium relationships used in this control method.

The total alkalinity of a synthetic water can be expressed as:

$$[Alk] = [HCO_3^-] + 2[CO_3^=] + [OH^-] - [H^+] \qquad (6)$$

$$= C_T(a_1 + 2a_2) + [OH^-] - [H^+] \qquad (7)$$

where [Alk] is the total alkalinity, eq/l, and

$$C_T = [H_2CO_3^*] + [HCO_3^-] + [CO_3^=]$$

$$a_0 = \frac{[H_2CO_3^*]}{C_T}$$

$$a_1 = \frac{[HCO_3^-]}{C_T}$$

$$a_2 = \frac{[CO_3^{-2}]}{C_T}$$

with all other concentrations expressed in mol/l. Considering the carbonate equilibrium listed in Table 5.2, Equation 6 can be transformed to

$$[Alk] = \frac{K_7 K_8 p_{CO_2}}{[H^+]} + \frac{2 K_7 K_8 p_{CO_2}}{[H^+]} + \frac{K_w}{[H^+]} - [H^+] \qquad (8)$$

where p_{CO_2} is the partial pressure of CO_2, atm, and K_w is the ion product of water.

If alkalinity is added in the form of $NaHCO_3$, Na_2CO_3 or $NaOH$ to a sample of distilled water, then an expression for the charge balance considering electroneutrality can be written:

$$[Na^+] + [H^+] = [HCO_3^-] + 2[CO_3^{-2}] + [OH^-] \qquad (9)$$

Therefore, [Alk] is identically equal to [Na^+]. Substituting [Na^+] for [Alk] in Equation 8, the following expression can be obtained, which shows the relationship between the partial pressure of CO_2 and pH for a given alkalinity.

$$p_{CO_2} = \frac{\{[Na^+][H^+] + [H^+]^2 - K_w\}[H^+]}{2 K_7 K_8 K_9 + K_7 K_8 [H^+]} \quad (10)$$

Equation 10 shows that the pH of an aqueous solution in a closed system can be controlled by varying the partial pressure of CO_2 in a mixture of carbon dioxide, nitrogen and oxygen. The solution will always maintain a constant partial pressure of oxygen. In the present study the final mixture of gases was analyzed occasionally prior to bubbling through the test solution to insure that it had a constant p_{O_2} and a desired p_{CO_2}, computed using Equation 10.

In a similar fashion one can develop a relationship to maintain a constant value of C_T at any pH:

$$[Na^+] = \frac{C_T \{2 K_7 K_8 K_9 + K_7 K_8 [H^+]\}}{K_7 K_8 K_9 + K_7 K_8 [H^+] + K_7 [H^+]^2} - \frac{[H^+]^2 - K_w}{[H^+]} \quad (11)$$

The amount of base needed for a desired C_T at any given pH is calculated using Equation 11 and added in the form of $NaHCO_3$, Na_2CO_3 or NaOH. Then a mixture of carbon dioxide, nitrogen and oxygen is bubbled through the test solution until the desired pH is reached. At this point C_T should be at the desired level. It should be noted, however, that for any decrease in pH, the partial pressure of CO_2 will have to be adjusted and the requisite amount of base calculated from Equation 11 will have to be added. In this particular study, pH and total alkalinity instead of C_T was maintained at a constant level. The aforementioned equilibrium relationships at 25°C have been shown graphically in Figure 5.2 for a constant C_T and a constant alkalinity [22]. Similar graphs were obtained for other alkalinity levels.

Buffer Capacity

The buffer capacity of the test solutions was calculated using the relationship:

$$\beta = 2.3 \{[H^+] + [OH^-] + C_T [a_1 (a_0 + a_2) + 4 a_2 a_0]\} \quad (12)$$

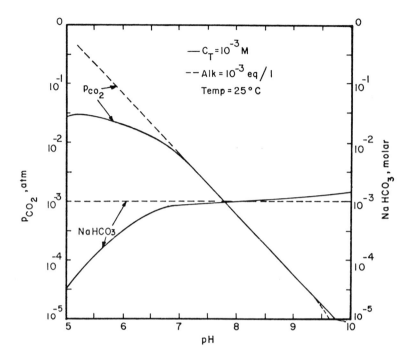

Figure 5.2. Relationship between p_{CO_2} and pH for constant C_T and alkalinity.

where β is the buffer capacity, eq/pH. Appropriate corrections for ionic strength and temperature were applied to all equilibrium constants used in the computation of β. It is apparent from Equations 7 and 12 that by keeping alkalinity and pH constant during an experiment the buffer capacity of the test solution could be maintained at a constant level. Figure 5.3 shows β values at different pH values.

Calculations for Supersaturation

The precipitation of $FeCO_3$ can be described by the reaction

$$FeCO_3(s) + H^+ = Fe^{2+} + HCO_3^- \qquad (13)$$

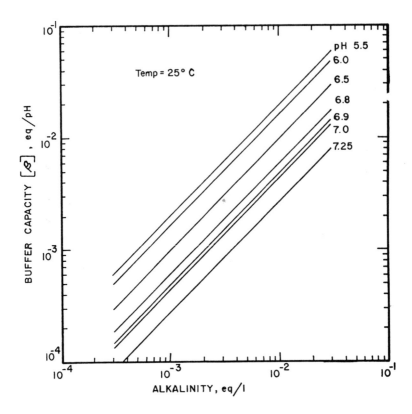

Figure 5.3. Relationship between β and alkalinity.

The reaction quotient for this reaction is given by

$$Q^c = \frac{[Fe^{2+}]_m [HCO_3^-]_m}{[H^+]_m} \tag{14}$$

where $[\ ]_m$ represents the measured concentrations of the various species. The equilibrium relationship for a test solution is given by

$$\frac{[Fe^{2+}][HCO_3^-]}{[H^+]} = K^c = \frac{K_4^c}{K_9^c} \tag{15}$$

the superscript c referring to values at a given ionic strength and the subscripts for K's referring to equation numbers in Table 5.2. For a test solution at equilibrium

$$Q^c = K^c \tag{16}$$

The equilibrium constant for the reaction at 25°C and zero ionic strength is given by

$$\frac{(Fe^{2+})(HCO_3^-)}{(H^+)} = K = \frac{K_4}{K_9} \tag{17}$$

Therefore, K is equal to K^c ($\gamma_{HCO_3^-}$) where γ represents the activity coefficient for single ions. The value of γ for different ions was determined using Davies' equation

$$-\log \gamma_i = A z_i^2 \left(\frac{\sqrt{I}}{1 + \sqrt{I}} - 0.3\,I \right) \tag{18}$$

where A is a constant having a value of 0.5 for water at 25°C, z is the charge of a given ion, and I is the ionic strength of the test solution.

Correcting for activity, the reaction quotient can be rewritten as

$$\log Q^a = \log [Fe^{2+}]_m + \log [HCO_3^-]_m - \log [H^+]_m +$$

$$\log \gamma_{Fe^{2+}} + \log \gamma_{HCO_3^-} - \log \gamma_{H^+} \tag{19}$$

where superscript a represents values corrected for activity. Simplifying

$$\log Q^a = \log Q^c + 0.5 \left(\frac{\sqrt{I}}{1 + \sqrt{I}} - 0.3\,I \right) \left(z_{H^+}^2 - z_{HCO_3^-}^2 - z_{Fe^{2+}}^2 \right)$$

$$= \log Q^c - 2 \left(\frac{\sqrt{I}}{1 + \sqrt{I}} - 0.3\,I \right) \tag{20}$$

At 25°C and zero supersaturation

$$\log Q^a = \log K = \log \frac{K_4}{K_9} = 0.085 \qquad (21)$$

using a value of 5.7×10^{-11} for K_4 as reported by Singer and Stumm instead of 2.1×10^{-11}, as listed in Table 5.2. The degree of supersaturation was computed from

$$S = \frac{Q^a}{K} \qquad (22)$$

EXPERIMENTAL METHOD

Batch studies were performed to determine the effect of buffer capacity on the oxygenation kinetics of Fe(II). Figure 5.4 shows a schematic diagram of experimental apparatus used. Each experiment was carried out in a reaction vessel containing 3.6 l of triple-distilled, deionized water. The triple-distilled water was obtained using a borosilicate glass still. A desired amount of

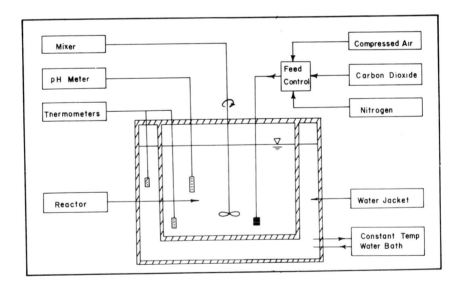

Figure 5.4. Schematic diagram of experimental apparatus.

alkalinity was added as Na_2CO_3 or $NaHCO_3$, and the solution was equilibrated to a temperature of 25 ± 0.5°C. Bicarbonate-CO_2 buffers were prepared by bubbling a predetermined mixture of carbon dioxide, nitrogen and oxygen until a desired pH level was reached. The detailed method for the control of pH and alkalinity has been described previously.

After the attainment of equilibrium with respect to pH, alkalinity, and temperature, a predetermined amount of a freshly prepared stock solution of ferrous perchlorate, prepared by dissolving $Fe(ClO_4)_2$ in 0.1 M $HClO_4$ and equilibrated to the same gas mixture, was added to the bicarbonate solution and a stopwatch was started. During the entire period of pH adjustment and the oxidation reaction the solution was well stirred with glass impellers. The concentration of Fe(II) added and the alkalinity of the test solution were carefully selected so that the degree of oversaturation with respect to $FeCO_3$ was negligible. The rate of oxidation was measured by analyzing reactor samples collected at various times for total Fe(II) iron.

Continuous flow studies were performed on the groundwater from Old Town, Maine. The detention time in the reactor was varied from 20 to 225 minutes. The inflow and outflow from the reactor were metered and controlled with a rotameter so that the desired residence time in the vessel could be obtained. Initially, the oxygenation rate was determined by doing replicate batch experiments using the natural groundwater.

Aliquots collected from the reactor at various intervals were added to 1.5 ml of concentrated $HClO_4$ in order to quench the reaction. The samples were then analyzed for total Fe(II) using the colorimetric reagent bathophenanthroline in accordance with the method outlined by Lee and Stumm [16] and later modified by Ghosh, et al. [17]. In some experiments both filterable Fe(II) and total Fe(II) were determined on each sample. Although in all experiments there was a slight initial supersaturation (S_o was never larger than 2.5), there was no discernible precipitation of $FeCO_3(s)$ as evidenced by identical values for the measured concentrations of total and filterable Fe(II). Therefore, in later experiments no measurement of filterable Fe(II) was made. All other analytical determinations were performed according to the 13th edition of *Standard Methods*.

RESULTS AND DISCUSSION

Buffer Capacity Effects

Experiments conducted at a given alkalinity and pH indicated that the oxygenation of Fe(II) in bicarbonate waters conforms to the kinetics proposed by Stumm and Lee [4]. The plot of log [Fe(II)] versus time at a given alkalinity, partial pressure of oxygen, and pH yields a straight line as shown in Figure 5.5, which indicates a first-order reaction. The slope of the line gives the value

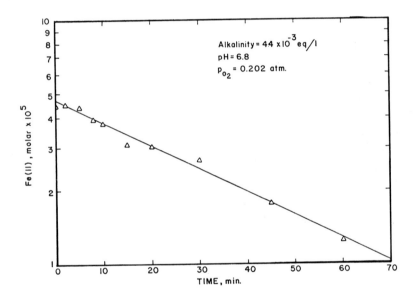

Figure 5.5. Oxygenation of iron(II).

of K, the overall first order reaction rate constant, as

$$K = k\,[OH^-]^2\,[p_{O_2}] \qquad (23)$$

where k is the oxygenation rate constant in the kinetic model proposed by Stumm and Lee [4] expressed in $l^2\,mol^{-2}\,atm^{-1}\,min^{-1}$, or

$$-\frac{d\,[Fe(II)]}{dt} = k\,[Fe(II)]\,[p_{O_2}]\,[OH^-]^2 \qquad (24)$$

Substituting molar concentration of dissolved oxygen in place of p_{O_2}, a new value may be obtained for the rate constant, k_m, expressed in $l^3 \, mol^{-3} \, min^{-1}$. The values of k_m reported by Stumm and Lee [4] and Morgan and Birkner [18] are shown in Table 5.3. The oxygenation rate equation can be rewritten in a convenient form,

$$-\frac{d \ln [Fe(II)]}{dt} = \frac{k_H [O_2 \, (aq)]}{[H^+]^2} \qquad (25)$$

where k_H is the rate constant expressed in $min^{-1} \, mol \, l^{-1}$. Table 5.3 shows a comparison of rate-constant values obtained in the present study with those reported in the literature, which were corrected for temperature using an activation energy of 23 kcal. All these values are for a temperature of 25°C and alkalinities ranging from 2.8×10^{-2} to 4.2×10^{-2} eq/l. It can be seen that the k values obtained in this study compare well with reported values.

The dependence of the reaction rate K on buffer capacity for systems maintained at a constant pH of 6.8, a temperature of 25°C, and at a p_{O_2} of 0.2 atm is shown in Figure 5.6. It is apparent that higher rates of oxygenation are associated with higher alkalinities, that is, higher buffer capacities. The variation in the values of k and K for different values of buffer capacity, a pH of 6.8, p_{O_2} of 0.2 atm, and a temperature of 25°C are shown in Table 5.4. The buffer capacity of a solution depends on its pH, alkalinity, temperature, and ionic strength.

A bivariate linear regression analysis was performed on the buffer capacity values and the corresponding values of K, the overall first-order rate constant. For values of β ranging between 1.0×10^{-4} and 4.0×10^{-3} eq/pH, the effect of β on the overall rate constant is insignificant. However, at values of β higher than 4.0×10^{-3} eq/pH, there is a definite correlation between these two variables. The correlation coefficient for the regression analysis was found to be 0.98, indicating that within the range studied the buffer capacity affects the Fe(II) oxygenation rate. Therefore, the term β should be included in the Fe(II) oxygenation rate equation

$$-\frac{d [Fe(II)]}{dt} = k \, [Fe(II)] \, [p_{O_2}] \, [OH^-]^2 \, [\beta]^n \qquad (26)$$

Table 5.3
Comparison of Fe(II) Oxygenation Rate Constants at 25°C

Alkalinity eq/l	k $\dfrac{liter^2\ mol^{-2}}{atm^{-1}\ min^{-1}}$	k_m $\dfrac{liter^3\ mol^{-3}}{min^{-1}}$	k_H $\dfrac{min^{-1}\ mol}{liter^{-1}}$	Reference
$2.8 \times 10^{-2} - 4.07 \times 10^{-2}$	$(5.0 \pm 0.4) \times 10^{13}$	3.8×10^{16}	3.8×10^{-2}	Present Study
$2.9 \times 10^{-2} - 3.9 \times 10^{-2}$	$(14 \pm 5) \times 10^{13}$	$(10.8 \pm 4.5) \times 10^{16}$	5.4×10^{-12}	4
$1.8 \times 10^{-2} - 3.2 \times 10^{-2}$	$2.5 \times 10^{13} - 8.8 \times 10^{13}$	$2 \times 10^{16} - 7 \times 10^{16}$		18

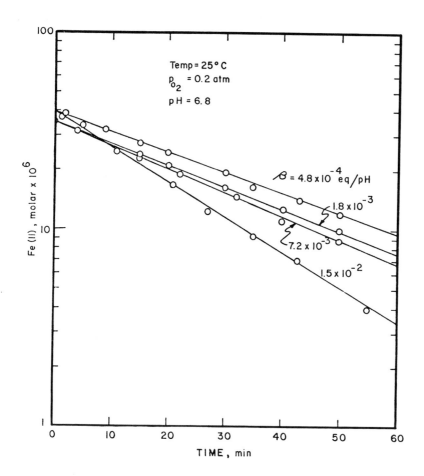

Figure 5.6. Oxygenation of Fe(II) at different buffer capacities.

The new overall first order rate constant is

$$K = k\,[p_{O_2}]\,[OH^-]^2\,[\beta]^n \tag{27}$$

In the present study, since temperature, pH and p_{O_2} were held constant, K should be directly related to β. Therefore,

$$k\,[OH^-]^2\,[p_{O_2}] = C \tag{28}$$

210 Aqueous-Environmental Chemistry of Metals

Table 5.4

Variations of k and K with Alkalinity and Buffer Capacity

Alkalinity eq/l	Buffer Capacity eq/pH	$\dfrac{10^{-13} \; k}{l^2 \; mol^{-2} \; atm^{-1} \; min^{-1}}$	$\dfrac{K^a}{min^{-1}}$
3.0×10^{-4}	1.32×10^{-4}	2.80	0.02231
1.12×10^{-3}	0.48×10^{-3}	3.07	0.02438
4.4×10^{-3}	1.84×10^{-3}	2.73	0.02166
6.0×10^{-3}	2.48×10^{-3}	3.45	0.02659
1.2×10^{-2}	0.48×10^{-2}	3.21	0.02549
1.8×10^{-2}	0.72×10^{-2}	3.96	0.03150
2.8×10^{-2}	1.07×10^{-2}	4.73	0.03759
4.07×10^{-2}	1.52×10^{-2}	5.29	0.04200

$^a K = k \, [p_{O_2}] \, [OH^-]^2$

where C is a constant and

$$K = [\beta]^n \, [C] \qquad (29)$$

A plot of log K versus log β yields a straight line as shown in Figure 5.7.

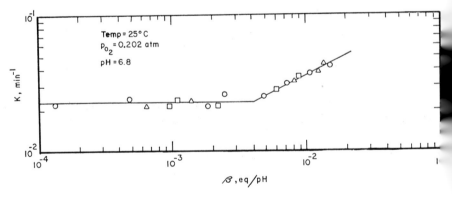

Figure 5.7. Relationship between K and β.

The value of n estimated from the slope of line was found to be 0.50. It is reemphasized that the effect of β becomes significant only at values in excess of 4.0×10^{-3} eq/pH, which corresponds to an alkalinity of approximately 340 mg/l as $CaCO_3$ at a pH of 6.8. Therefore, in most natural waters the influence of β on the oxygenation rate will be insignificant. The apparent lack of dependence of oxygenation rate on β at low values is probably due to the slow response of the bicarbonate buffer system to localize acidity changes induced by the oxygenation reaction as proposed by Stumm and Lee [4].

The new rate law for highly buffered bicarbonate solutions can be written as

$$-\frac{d[Fe(II)]}{dt} = k[Fe(II)][OH^-]^2[p_{O_2}][\beta]^{\frac{1}{2}} \quad (30)$$

where k is expressed in l^2 atm^{-1} mol^{-2} min^{-1} eq^{-1} pH. The value of k was found to be 1.3×10^{15} at 25°C for values of β in the range of 4.0×10^{-3} to 1.5×10^{-2} eq/pH and pH ranging from 5.5 to 7.2.

Experiments were conducted at different levels of pH and β. Figure 5.8 shows a plot of the logarithm of the overall rate

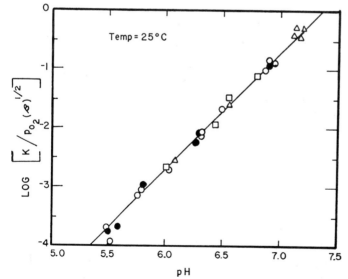

Figure 5.8. Relationship between oxygenation rate and pH.

constants divided by p_{O_2} and $\beta^{1/2}$ as a function of pH. The resultant line through the experimental data points has a slope of 2, indicating a second-order relationship between the overall reaction rate K and the hydroxide ion concentration in congruence with the kinetic models proposed by previous investigators [2,4].

Supersaturation Effects

In an attempt to explain the observed increase in the Fe(II) oxygenation rate with an increase in buffer capacity, various possibilities were investigated. In a water with high alkalinity, the solubility of Fe(II) is primarily influenced by $FeCO_3(s)$ in the pH range of 6 to 10 as indicated in Figure 5.1. In the pH range investigated in this research, a water buffered with the carbonate system could become oversaturated with $FeCO_3(s)$ causing a portion of Fe(II) to precipitate. In order for $FeCO_3(s)$ to precipitate "an energy barrier must be overcome before crystallization can occur" [8].

Precipitation will occur only when the degree of supersaturation S, the product of the activities of the ions divided by the solubility product constant, exceeds a critical value. The degree of initial supersaturation S_o was computed for certain experimental solutions as listed in Table 5.5. In computing these values the solubility product constant for $FeCO_3(s)$ was taken as 5.7×10^{-11}, as reported by Singer and Stumm [19]. It can be seen from Table 5.5 that the initial supersaturation varied from 0.041 to 2.17. Although all of the test solutions listed in the table appeared to be supersaturated, the values of S_o were too small to initiate precipitation of crystalline $FeCO_3(s)$.

Applicability of CSTR Theory

Fluid flow models have been used extensively to determine the effects of the flow pattern in a reactor on the rate of chemical conversion. In this research, an attempt was made to determine if the batch kinetics data on Fe(II) oxygenation could be used to predict the efficiency of the chemical conversion of Fe(II) to Fe(III) in a continuous stirred-tank reactor. The theory of CSTR,

Table 5.5

Initial Supersaturation of Experimental Solutions at 25°C and a pH of 6.8

Alkalinity $(eq/1 \times 10^4)$	C_T $(M \times 10^4)$	$[Fe^{2+}]m$ $(M \times 10^5)$	$[HCO_3^-]m$ $(M \times 10^4)$	$[H^+]m$ $(M \times 10^7)$	Q^c	$\log Q^c$	I $(M \times 10^5)$	Activity[a] Correction	$\log Q^a$	Q^a	S_o[b]
3.0	4.1	5.64	1.5	1.59	0.053	-1.276	22.5	0.029	-1.305	4.96×10^{-2}	0.041
11.2	15.1	1.22	5.6	1.59	0.043	-1.367	84.0	0.051	-1.418	3.82×10^{-2}	0.031
44.0	59.5	4.73	22.0	1.59	0.65	-0.187	330.0	0.107	-0.294	5.08×10^{-1}	0.416
60.0	81.1	3.49	30.0	2.59	0.66	-0.181	450.0	0.123	-0.304	4.98×10^{-1}	0.408
120.0	162.0	3.31	60.0	1.59	1.25	+0.097	900.0	0.168	-0.071	8.49×10^{-1}	0.696
182.0	243.0	2.86	91.0	1.59	1.64	+0.215	1215	0.191	+0.024	1.06	0.869
280.0	378.0	3.40	140.0	1.59	2.99	+0.476	2100	0.241	+0.235	1.72	1.41
407.0	550.0	3.94	203.5	1.59	5.04	+0.702	3053	0.279	+0.423	2.65	2.17

[a] $\log Q^a = \log Q^c - 2 \left(\dfrac{\sqrt{I}}{1+\sqrt{I}} - 0.3\, I \right)$

[b] $S_o = \dfrac{Q^a}{(K_4/K_9)} = \dfrac{Q^a}{1.22}$

Temperature corrections for equilibrium constants were made using the van't Hoff temperature relationship.

or back mix reactors, has been discussed extensively in the literature [20]. At steady state, a mass balance around the reactor gives

$$\frac{C_o}{C} - 1 = K \theta_m \qquad (31)$$

where

C_o = concentration of Fe(II) entering CSTR
C = concentration of Fe(II) leaving CSTR
K = overall first order rate of oxygenation
θ_m = mean hydraulic residence time in the reactor.

The mean residence time was varied between 20 to 225 minutes and was determined at each flow rate by tracer studies. Oxygenation experiments were started only after the reactor attained steady state.

In this phase of the study the raw well water from Old Town, Maine, was used as the test solution. The operating characteristics of the CSTR tests are listed in Table 5.6. The rate constant for

Table 5.6

Operating Characteristics of CSTR Tests on Groundwater from Old Town, Maine

Characteristic	Result
Initial Fe(II) concentration	3.87×10^{-5} M
Alkalinity	1.4×10^{-3} M
pH	6.8
Partial pressure of O_2 (p_{O_2})	0.181 atm
Buffer capacity (β)[a]	8.37×10^{-4} eq/pH
Total organic carbon (as C)	0.086×10^{-3} M
Oxygenation rate constant (k)[a]	2.87×10^{13} l^2 mol^{-2} atm^{-1} min^{-1}
Overall oxygenation rate (K)	0.0205 min^{-1}
Initial supersaturation (S_o)	0.019

[a]It was assumed that for such a small value of β, there was no appreciable effect of buffer capacity on k.

this natural water was in excellent agreement with those obtained with synthetic waters as shown in Table 5.6. The effect of oversaturation was neglected as the initial oversaturation was less than 0.1. Figure 5.9 shows a comparison between theoretical and experimental oxygenation efficiencies. The observed efficiency was

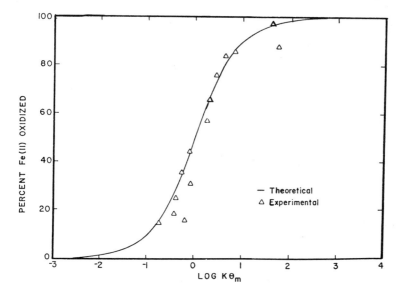

Figure 5.9. Oxygenation of Fe(II) in a continuous stirred treatment reactor.

less than the theoretical value for most experimental runs, indicating that process designs based on a direct scale-up of batch kinetics data will tend to overestimate the performance of a prototype. Similar observations were made by other researchers [21]. However, CSTR studies can be used to yield results that are meaningful and successfully applicable to prototype designs.

CONCLUSIONS

This research has shown that the buffer capacity of a water, at values higher than 4.0×10^{-3} eq/l, has a definite influence on the oxygenation of Fe(II) in bicarbonate solutions. A new Fe(II) oxygenation rate law has been proposed for highly buffered bicarbonate solutions. The value of k was 1.3×10^{15} l^2 atm^{-1}

mol^{-2} min^{-1} eq^{-1} pH for solutions with buffer capacities ranging from 4.0×10^{-3} to 1.5×10^{-2} eq/pH. In most natural waters having alkalinities less than 300 mg/l as $CaCO_3$, the oxygenation reaction may proceed at a slow rate because of poor buffering conditions.

The efficiency of iron removal for a given water can be predicted reasonably well by applying CSTR theory to the rate constant obtained from batch oxidation studies in the laboratory. Optimal design criteria for the deferrization process can be developed from batch and continuous flow studies.

ACKNOWLEDGMENT

This study was supported in part by a grant from the Maine Water Utilities Association. The author wishes to thank Robert Jobin and Russell Boyd for technical assistance.

REFERENCES

1. Applebaum, S. B. *Water & Sew. Works, 103,* 258 (1956).
2. Just, G. *Z. Phys. Chem., 63,* 385 (1908).
3. Holluta, J. and M. Eberhardt. *Vom Wasser, 24,* 79 (1957).
4. Stumm, W. and G. F. Lee. *Ind. Eng. Chem., 53,* 143 (1961).
5. Singer, P. C. and W. Stumm. *Proc. Conference on Acid Mine Drainage Research.* (1968), p. 12.
6. Longley, J. M. M. S. Thesis, University of Illinois (1961).
7. Ghosh, M. M., J. T. O'Connor, and R. S. Engelbrecht. *J. San. Eng., Div., ASCE, 92,* 120 (1966).
8. Stumm, W. and P. C. Singer. *J. San. Eng. Div., ASCE, 92,* 120 (1966).
9. Oldham, W. K. and E. F. Gloyna. *J. Amer. Water Works Assoc., 61,* 610 (1969).
10. Leussing, D. L. and I. M. Kolthoff. *J. Amer. Chem. Soc., 75,* 2476 (1953).
11. Gayer, K. H. and L. Woontner. *J. Phys. Chem., 60,* 1509 (1956).
12. Latimer, W. E. *Oxidation Potentials,* 2nd ed. (Englewood Cliffs, N.J.: Prentice Hall, 1952), p. 222.
13. Ellis, A. J. *Amer. J. Sci., 257,* 217 (1959).
14. Harned, H. S. and R. Davies, Jr. *J. Am. Chem. Soc., 65,* 2030 (1943).
15. Harned, H. S. and S. R. Scholes. *J. Am. Chem. Soc., 63,* 1706 (1941).
16. Lee, G. F. and W. Stumm. *J. Amer. Water Works Assoc., 52,* 1567 (1960).
17. Ghosh, M. M., J. T. O'Connor, and R. S. Engelbrecht. *J. Amer. Water Works Assoc., 59,* 897 (1967).

18. Morgan, J. J. and F. B. Birkner. *J. San. Eng. Div., ASCE, 92,* 137 (1966).
19. Singer, P. C. and W. Stumm. *J. Amer. Water Works Assoc., 62,* 198 (1970).
20. Levenspiel, O. *Chemical Reactor Engineering.* (New York: John Wiley and Sons, 1968).
21. O'Melia, C. R. "Oxygenation of Iron(II) in Continuous Reactors," North Carolina Water Resources Inst. Report No. 23 (1969).
22. Roberts, R. F. and H. E. Allen. *Trans. Amer. Fisheries Soc., 101,* 752 (1972).

CHAPTER 6

CRYSTAL GROWTH KINETICS OF MINERALS ENCOUNTERED IN WATER TREATMENT PROCESSES

George H. Nancollas
　Chemistry Department
　State University of New York at Buffalo
　Buffalo, New York

Michael M. Reddy
　New York State Health Department
　Division of Laboratories and Research
　New Scotland Avenue
　Albany, New York

Introduction	219
Results and Discussion	225
Calcium Carbonate Studies	225
Magnesium Hydroxide Studies	239
Dicalcium Phosphate Studies	244
Conclusion	250

INTRODUCTION

Heterogeneous reactions involving aqueous solutions and minerals play an important role in natural water systems and in water treatment processes. Recent attempts to describe such reactions using an equilibrium model have had varying degrees of success [1,2]. Pronounced discrepancies between the observed and the calculated equilibrium composition of the natural waters show that a kinetic modeling of these systems would be much more appropriate than simply considering them to be in thermodynamic equilibrium. Furthermore, failure of many heterogeneous water treatment processes to reach equilibrium demonstrates the desirability of a kinetic analysis. This is especially true in applications involving seeded growth [3] and in the implementation of water treatment processes designed to eliminate the formation of sludges [4,5,6]. Kinetic analysis of mineral reactions encountered in water treatment processes will facilitate modification of existing processes in accordance with environmental protection policies, enabling better design optimization and decreased capital costs.

In order to be able to make a detailed kinetic analysis of crystal growth from aqueous solution, it is necessary to know the equilibrium distribution of species in the solutions. Complex formation can readily be taken into account in a thermodynamic analysis of the solution phase by employing the thermodynamic ion association constants given in Table 6.1 [7]. Activity

Table 6.1

Thermodynamic Ion Association Constants at $25°C$

Species	K, Ion Association Constant $l.\ mol^{-1}$	Reference
$MgHCO_3^+$	13.1	[24]
$CaHCO_3^+$	19.4	[24]
$CaCO_3^°$	1.59×10^3	[25]
$CaSO_4^°$	169.	[20]
$CaPO_4^-$	2.9×10^6	[26]
$CaHPO_4^°$	548.	[26]
$CaH_2PO_4^+$	25.6	[26]
$MgHPO_4^°$	316.	[27]
$CaOH^+$	25.	[7]
$MgOH^+$	355.	[15]

coefficients in the solutions are conveniently calculated at ionic strengths below 0.1 M by using an extended form of the Debye-Hückel equation such as that proposed by C. W. Davies [8]*

$$\log f_z = -0.5\, z^2\, [I^{1/2}/(1 + I^{1/2}) - 0.3I] \qquad (1)$$

Thermodynamic solubility products, K_{sp}, for several important minerals encountered in water treatment processes are given in Table 6.2. The use of such data requires that the solutions be in equilibrium with a thermodynamically stable solid phase.

Table 6.2

Thermodynamic Solubility Products at 25°C

Salt	K_{sp}, Solubility Product	Reference
$CaCO_3$	$4.01 \times 10^{-9}\, mol^2\, l^{-2}$	[28]
$Mg(OH)_2$	$1.38 \times 10^{-11}\, mol^3\, l^{-3}$	[15]
$CaHPO_4 \cdot 2H_2O$	$2.1 \times 10^{-7}\, mol^2\, l^{-2}$	[23]
$CaSO_4 \cdot 2H_2O$	$2.58 \times 10^{-5}\, mol^2\, l^{-2}$	[20]

The thermodynamic driving force for a crystallization reaction can be expressed in terms of the solution supersaturation, S, in Figure 6.1. For simple salts, S is defined as the ratio

$$S = (m - m_o)/m_o \qquad (2)$$

where m is the concentration of the supersaturated solution and m_o is the saturation value, given by

$$m_o^2 = K_{sp}/f_z^2 \qquad (3)$$

---

*See page 251 for explanation of notations used in this chapter.

222 *Aqueous-Environmental Chemistry of Metals*

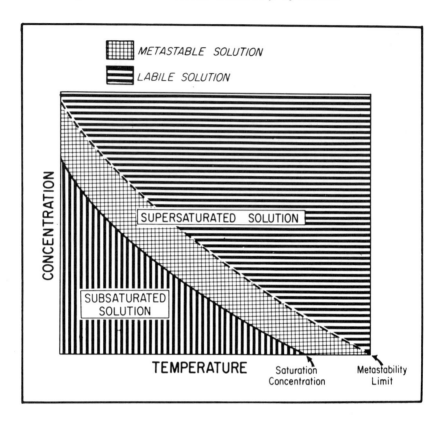

Figure 6.1. The solubility diagram showing subsaturated, labile and metastable supersaturated solutions.

A supersaturated solution may be either labile, undergoing spontaneous precipitation through nucleation and subsequent growth in a short period of time, or metastable, in which case no observable precipitate is formed in a relatively long period. The metastable region is depicted in Figure 6.1. Its extent varies from salt to salt and in some instances the critical concentration above which spontaneous precipitation takes place can be as much as ten times the solubility value. An illustration of the processes involved in crystallization from each type of supersaturated solution is presented in Figure 6.2. Here crystal growth is initiated in the metastable supersaturated solution by inoculation with added seed crystals.

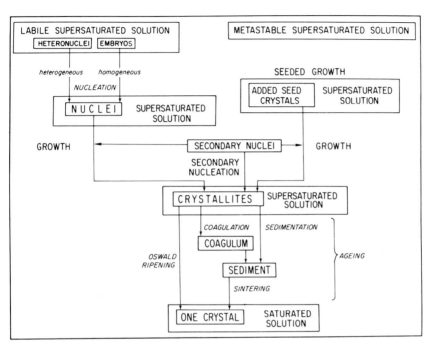

Figure 6.2. Processes involved in crystal growth and nucleation from supersaturated solution.

In the absence of secondary nucleation, the growth of seed crystals from metastable supersaturated solutions has been found for many salts to follow a rate law of the form [9]:

$$\text{rate of crystal growth} = -dm/dt = k\, s(m-m_o)^n \qquad (4)$$

where m is the concentration of species A in solution at time t, dm/dt the rate of crystal growth, k the crystal growth rate constant, s the crystal surface area available for growth, and n a constant. Kinetic investigations of crystal growth for a number of sparingly soluble salts have yielded experimental rate laws similar to Equation 4, and this can be interpreted in terms of a crystal growth mechanism involving a rate-controlling surface reaction. A typical self-propagating step on a crystal surface is shown in Figure 6.3 in which the overall growth process consists

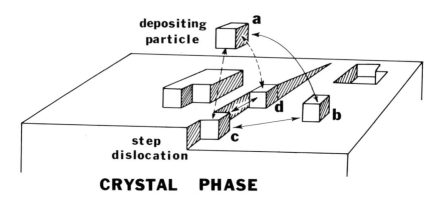

Figure 6.3. Steps involved in crystal growth.

of (1) diffusion of the depositing molecule or ion in the bulk of the solution (position a) up to the crystal surface where it is (2) adsorbed (position b). Thereafter, (3) a surface diffusion process to a step edge (position c) offers additional stabilization since in position c, two surfaces of the particle are in contact with the crystal faces (4). A further one-dimensional diffusion along the step to a kink site (position d) enables the molecule or ion to be incorporated into the crystal lattice. Much experimental evidence [9] points to a surface reaction such as (2), (3) or (4) as the rate-determining step rather than bulk diffusion, reaction (1), of material to the growing crystal surface.

Crystal growth studies of a number of minerals encountered in water treatment processes have been made using a highly reproducible seeded crystal growth technique. This method has a number of advantages as compared with the numerous spontaneous precipitation studies that have been conducted with labile supersaturated solutions. In spontaneous precipitation experiments, it is difficult to distinguish between crystal nucleation and subsequent crystal growth and to separate the effects of heterogeneous and homogeneous nucleation (see Figure 6.2). The seeded growth technique enables not only reproducible measurement of crystal growth rates but also quantitative study of the effects of additives in modifying these rates.

In the seeded crystal growth experiments to be described in this paper, stable supersaturated solutions have been prepared either by the careful mixing of solutions containing the lattice ions of the salt to be precipitated, by changing the temperature of a saturated solution, or by the dissolution of a thermodynamically unstable form of the salt. Crystal growth is initiated in the stable supersaturated solution at the beginning of each run by inoculation with seed crystals. Seed crystals are prepared by spontaneous precipitation under conditions known to yield crystals of regular size and morphology, followed by repeated washing and aging for several months before use. The crystals are characterized by chemical analysis, by optical and electron microscopy, by X-ray powder diffraction and by the measurement of the crystal surface area using gas adsorption and isotopic exchange. Following inoculation with seed crystals, growth normally commences immediately and is accompanied by a decrease in the solution concentration of the precipitating ions. Concentration changes in solution can be measured directly by potentiometric or conductance measurements, or aliquots can be removed for analysis at known times. Atomic absorption analysis, complexometric titration, isotope dilution and spectrophotometry have all been used for such analyses.

RESULTS AND DISCUSSION

In this section the results of a number of studies of calcite seeded growth are discussed especially in relation to their applicability in water treatment processes. Simple seeded growth experiments, in both the presence and absence of additives, are examined together with the effect of surface area on the calcite crystal growth. The crystal growth of two other important minerals, magnesium hydroxide and dicalcium phosphate dihydrate, is also discussed.

Calcium Carbonate Studies

Crystal growth of calcium carbonate, a mineral commonly encountered in water treatment processes, has been characterized using a seeded growth procedure [10]. Typical experimental results for crystal growth from supersaturated solution on to calcite seed crystal are shown in Figure 6.4 (blackened circles).

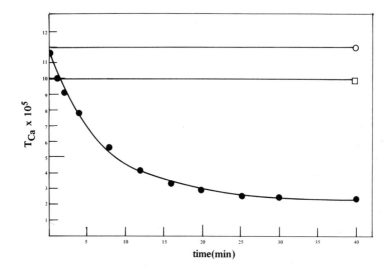

Figure 6.4. Plots of total calcium concentration as a function of time for calcium carbonate crystal growth. Stirring rate 385 rpm, 45.8 mgs seed B III per 100 ml of solution, temperature 25°C, total carbonate concentration 10^{-2} M, pH = 9.1. ● Expt. 32, 0 ppm NTPP; □ Expt. 34B, 2.5 ppm NTPP; ○ Expt. 33A, 10 ppm NTPP.

These data follow a rate law of the form [10]:

$$\frac{-d[T_{Ca}]}{dt} = k_s (T_{Ca} - T_{Ca}^\circ)^2 \qquad (5)$$

Analysis of calcite crystal growth data is facilitated by employing the integrated form of Equation 5:

$$(T_{Ca} - T_{Ca}^\circ)^{-1} - (T_{Ca}^i - T_{Ca}^\circ)^{-1} = k_s t \qquad (6)$$

In the integration of Equation 5 the equilibrium total calcium molarity, T_{Ca}°, has been treated as a constant. Maximum variation

Crystal Growth Kinetics of Minerals 227

in T_{Ca}° during a growth run was approximately 10%. In previous investigations of calcium carbonate crystal growth, an alternative rate expression was used to describe the growth rate when the equilibrium calcium ion concentration and carbonate ion concentration changed appreciably during growth [11,12]. The equilibrium calcium ion concentration is calculated using the thermodynamic solubility product of calcite at 25°C. The association constants for calcium carbonate, calcium bicarbonate and calcium hydroxide ion pairs are given in Table 6.1. Table 6.3 presents data for calcite growth and Figure 6.5 is a plot of the integrated rate law for the data used to prepare Table 6.3. These data illustrate the experimental reproducibility of the growth experiments and the agreement of the experimental data with the integrated rate law given in Equation 6.

Table 6.3

Crystallization of Calcite from Supersaturated Solution*

Expt. No.	Total Concentrations		pH_O	$[Ca^{2+}]_i$ 10^5 M	$[CO_3^{2-}]_i$ 10^4 M	k $l\ mol^{-1}\ min^{-1}$ mg seed/ 100 ml
	Calcium 10^5 M	Carbonate 10^2 M				
32	11.704	1.000	9.155	6.966	8.458	57.6
32A	11.941	1.000	9.158	7.086	8.516	57.6
32B	11.945	1.000	9.158	7.087	8.511	57.6
34A II	9.958	1.000	9.060	6.293	6.898	57.6

*Stirring rate 385 rpm, 45.8 mgs seed per 100 ml of solution, temperature 25°C.

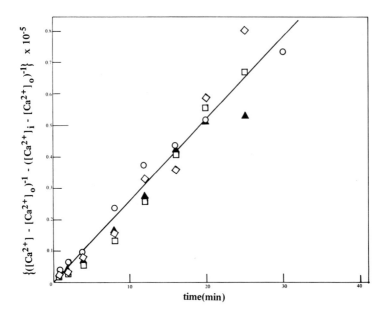

Figure 6.5. Plots of $(T_{Ca} - T_{Ca}^\circ)^{-1} - (T_{Ca}^i - T_{Ca}^\circ)^{-1}$. against time for calcium carbonate growth. Reproduced with permission from *Desalination* [10].

Calcite crystal growth has been studied in the presence of phosphorus-containing anionic additives at part per million (ppm) concentration levels. These anions are commonly employed in consumer and industrial products and may be expected to be encountered in increasing amounts in water supplies and water treatment processes. Polyphosphates, for example, are frequently used in water treatment procedures as scale control agents. The effect of sodium tripolyphosphate, at 2.5 and 10 ppm, on calcite seeded growth at 25°C is shown by the open symbols in Figure 6.4 [13], which clearly illustrates the complete inhibition of crystallization in the presence of low polyphosphate concentrations. The results of experiments employing higher seed concentrations and lower polyphosphate levels show that polyphosphate concentrations as low as 0.2 ppm have a significant effect on the calcite growth rate (Figure 6.6). These data are plotted according to the integrated rate law, Equation 6, for calcite seeded growth,

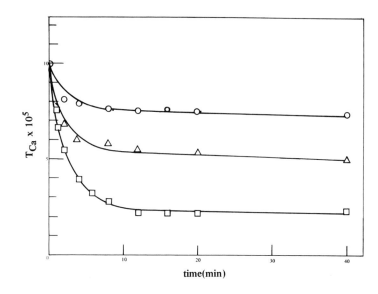

Figure 6.6. Plots of total calcium concentration against time for calcium carbonate crystal growth. Stirring rate 385 rpm, 95 mgs seed 8/5 per 100 ml of solution, temperature 25°C, total carbonate concentration, 10^{-2} M, pH = 9.1. □ Expt. 40, 0 ppm NTPP; △ Expt. 42, 0.2 ppm NTPP; ○ Expt. 41, 0.5 ppm NTPP

in Figure 6.7. Although the resulting linear plots in Figure 6.7 in the presence of polyphosphate do not pass through the origin, the small values for the slopes show the marked growth-inhibiting influence of this additive. Raistrick [14] has suggested that calcite growth inhibition results from a matching of calcite lattice parameters in the 111 face with the interatomic coordinates for the polyphosphate molecule, which is adsorbed on the surface. The seed crystals used in our experiments have a morphology characteristic of 100 crystal faces, as is clearly shown by the scanning electron micrographs in Figure 6.8. The observed marked inhibition of growth of the 100 calcite crystal faces by polyphosphate appears to result from a nonspecific adsorption of polyphosphate ion onto the crystal surface.

Calcium carbonate crystal growth experiments were also conducted in the presence of several phosphonate ions [10]. Phosphonate salts have been used commercially for scale control

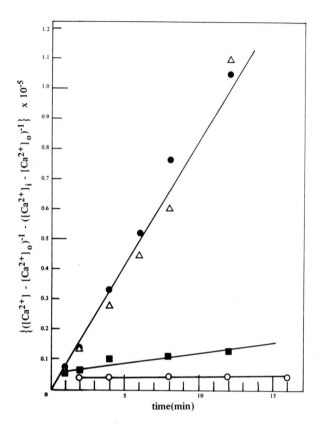

Figure 6.7. Plots of $(T_{Ca} - T_{Ca}^{\circ})^{-1} - (T_{Ca}^{i} - T_{Ca}^{\circ})^{-1}$ against time for calcium carbonate crystal growth. ● Expt. 39 and △ Expt. 40, 0 ppm NTPP; ■ Expt. 42, 0.2 ppm NTPP; ○ Expt. 41, 0.5 ppm NTPP.

in oil fields and in desalination processes, and they have considerable potential in the industrial field. Their major advantage over conventional scale inhibitors such as polyphosphates is their remarkable thermal and chemical stability. Experimental data obtained for calcium carbonate crystal growth in the presence of phosphonate additives is presented in Table 6.4. Total calcium ion concentration in solution during crystal growth in the presence of the highest additive concentrations employed in this study are shown in Figure 6.9. These data clearly demonstrate the marked calcium carbonate crystal growth inhibition in the presence of the phosphonates tested. Data shown

Figure 6.8. Scanning electron micrographs of calcite seed crystals employed in crystal growth experiments (x750); **A**, seed 8/5; **B**, seed B III.

Table 6.4
Crystallization of Calcite from Supersaturated Solutions in the Presence of Phosphonate Additives*

Expt. No.	Additive	Additive Concentration ppm	Additive Concentration mol/l × 10⁶	Total Concentrations Calcium 10⁵ M	Total Concentrations Carbonate 10² M	pH_0	$[Ca^{2+}]_0$ 10^5 M	$[CO_3^{2-}]_0$ 10^4 M	k $l\,mol^{-1}\,min^{-1}$ $mg\,seed/100\,ml$
34C	ENTMP	2.5	5.8	9.958	1.000	9.138	6.003	8.149	0.38
34E	TENTMP	2.5	5.1	9.958	1.000	9.159	5.908	8.525	0.29
34AA	ENTMP	0.5	1.2	9.958	1.000	9.155	5.933	8.449	1.31
34BB	TENTMP	0.5	1.0	9.958	1.000	9.155	5.929	8.462	0.87
34CC	HEDP	0.5	2.4	9.958	1.000	9.150	5.949	8.368	0.59
34DD	NTMP	0.5	1.6	9.958	1.000	9.149	5.954	8.352	1.55
34AAA	ENTMP	0.05	0.12	9.958	1.000	9.127	6.045	7.961	35.0
34AAB	ENTMP	0.12	0.28	9.958	1.000	9.162	5.889	8.588	9.0
34AAC	ENTMP	0.08	0.18	9.958	1.000	9.156	5.926	8.465	22.0

*Stirring rate 385 rpm, 45.8 mg of seed per 100 ml of solution, temperature 25°C.

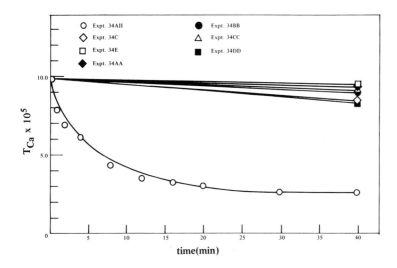

Figure 6.9. Plots of total calcium concentration against time for calcium carbonate crystal growth in the presence and absence of phosphonate ions. Reproduced with permission from *Desalination* [10].

in Figure 6.9 have been employed in the integrated rate law (Equation 6), and the results of this analysis are presented in Figure 6.10, which shows the marked decrease in the growth rate constant in the presence of phosphonates.

The results in Table 6.4 indicate that the calcite growth inhibition effectiveness of the various phosphonates in this program, at a fixed weight concentration of 0.5 ppm, is in the order EHDP > TENTMP > ENTMP > NTMP. The ratio of the rates of calcite crystal growth in the presence of 0.5 ppm TENTMP and 0.5 ppm ENTMP is equal within experimental error to the ratio of rates in the presence of 2.5 ppm TENTMP and ENTMP, suggesting that the order of inhibitor effectiveness observed at 0.5 ppm additive concentration may extend over a range of additive concentrations.

Calcium carbonate crystal growth has been examined over a wide range of ENTMP concentrations. The rate law integrated plots (Equation 6) for calcite crystal growth in the presence and absence of ENTMP are presented in Figure 6.11. It is clear that

234 Aqueous-Environmental Chemistry of Metals

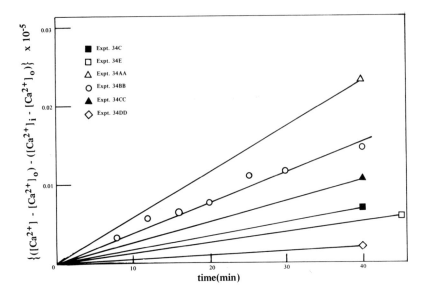

Figure 6.10. Plots of $(T_{Ca} - T_{Ca}^o)^{-1} - (T_{Ca}^i - T_{Ca}^o)^{-1}$ against time. Reproduced with permission from *Desalination* [10].

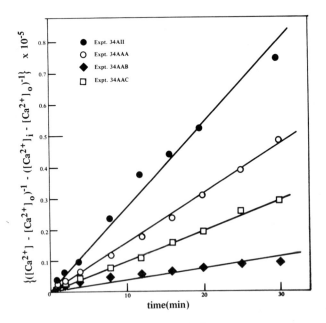

Figure 6.11. Plots of $(T_{Ca} - T_{Ca}^o)^{-1} - (T_{Ca}^i - T_{Ca}^o)^{-1}$ against time. Reproduced with permission from *Desalination* [10].

the rate law observed for calcite crystal growth in pure solutions is applicable to calcite crystal growth in solutions containing a range of phosphonate concentrations. The marked effect of ENTMP in reducing the calcium carbonate growth rate is illustrated in Figure 6.12, in which the rate constants for crystal

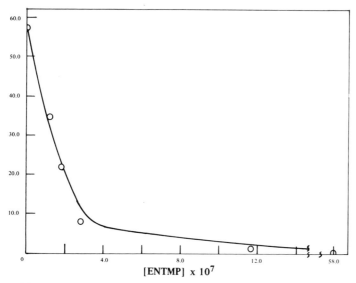

Figure 6.12. Plots of k, in the presence and absence of ENTMP, against [ENTMP]. Reproduced with permission from *Desalination* [10].

growth obtained from the slopes of the plots in Figure 6.11 (Table 6.4) are plotted as a function of ENTMP concentration. If the crystal growth inhibition by ENTMP is a result of simple adsorption of the additive molecules at the active growth sites on the crystal surfaces, then some form of adsorption isotherm should be applicable. In a growth experiment, suppose that a fraction a of the growth sites on the surface of the seed crystal are covered by adsorbed ions of molar concentration [ENTMP]. The desorption rate of these ions is $k_2 a$ and the adsorption rate may be written k_1 [ENTMP] $(1 - a)$, where k_2 and k_1 are the rate constants for desorption and adsorption, respectively. At equilibrium these rates are equal, and $a = k_1$ [ENTMP]$/(k_2 + k_1$ [ENTMP]). In terms of a simple Langmuir type adsorption $k = k_o (1 - a)$ where k_o is the crystal growth rate constant in the absence of ENTMP. Substituting for a:

$$k_0(k_0 - k)^{-1} = 1 + k_2(k_1[\text{ENTMP}])^{-1} \qquad (7)$$

In Figure 6.13, $k_0(k_0 - k)^{-1}$ is plotted against $[\text{ENTMP}]^{-1}$ and it is seen that the Langmuir isotherm satisfactorily describes the marked inhibiting effects of the ENTMP ions in terms of a monomolecular blocking layer of foreign ions at the growth sites on the crystal surface.

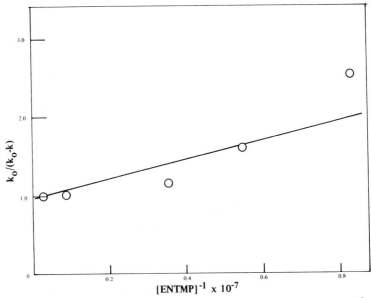

Figure 6.13. Langmuir isotherm plot of $k_0/(k_0-k)$ against $[\text{ENTMP}]^{-1}$, k_0, the calcite growth rate constant in the absence of ENTMP; k, the calcite growth rate constant in the presence of ENTMP. Reproduced with permission from *Desalination* [10].

The crystal surface area, especially of the added inoculating crystals, is another important parameter for the discussion of the growth rates in the seeded growth experiments. Changes in the methods of preparation of the seed crystals can produce marked differences in the effective surface area. Figure 6.14 shows scanning electron micrographs of three different calcite seed crystal preparations. The results of growth experiments conducted with each of these preparations, at various concentrations, are in

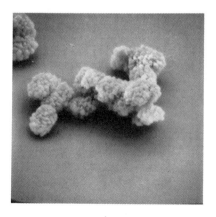

Figure 6.14. Scanning electron micrographs of calcite seed crystals employed in crystal growth experiments; **A**, seed B III (x525); **B**, seed B6P35 II (x525); **C**, seed B6P35 III (x2625).

Figure 6.15. The rates follow the integrated form of the kinetic Equation 6. In Table 6.5 the rate constant for each of these experiments is given together with surface areas determined by a dynamic nitrogen gas adsorption technique. It can be seen that the values of k, expressed per gram of seed material, differ markedly for the different seed preparations. Scanning electron micrographs (Figure 6.14) show that this may be due principally to variations in crystal surface area, and in Table 6.5 this seems to be the case. The agreement between the normalized growth rates constants (last column of Table 6.5), k'/A = 0.91 ± 0.21, is satisfactory. The result is of major importance in the application of seed crystal techniques to water treatment processes since the reaction rates may be altered markedly by a change in seed crystal surface area.

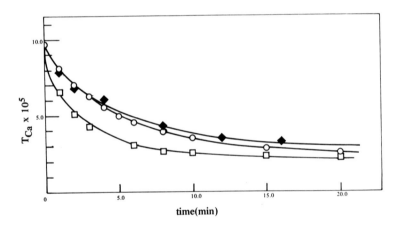

Figure 6.15. Plots of total calcium concentration against time for calcium carbonate crystal growth. ◆ Expt. 34A II; ○ Expt. 54; □ Expt. 55.

The results of these experiments illustrate the importance of low concentrations of phosphorus-containing ions and of seed crystal surface area in analyzing the calcite crystal growth rates. In the growth of other important modifications, such as dolomite, additional factors may play an equally important role. A seed crystal technique can be used to evaluate any potential modification of heterogeneous reactions used in water treatment processes at bench or pilot plant scale before changes are made in permanent plant facilities.

Crystal Growth Kinetics of Minerals

Table 6.5

Crystallization of Calcite from Supersaturated Solution*

	Experiment Number		
	34AII	54	55
Seed Crystal	I	II	III
Total Concentrations			
Calcium (10^5 M)	9.958	9.746	9.912
Carbonate (10^2 M)	1.000	0.906	0.902
pH_O	9.06	9.20	9.20
$[Ca^{2+}]_O$ (10^5 M)	6.293	5.767	5.884
$[CO_3^{2-}]_O$ (10^4 M)	6.898	8.426	8.370
k $\dfrac{1\ mol^{-1}\ min^{-1}}{mg\ seed/100\ ml}$	57.6	36.7	362
A $\dfrac{M^2}{gm}$	0.52	0.35	6.2
k'/A $\dfrac{1\ mol^{-1}\ min^{-1}}{M^2/liter}$	1.09	1.05	0.59

*Stirring rate 385 rpm, temperature 25°C.

Magnesium Hydroxide Studies

The kinetics of crystallization of magnesium hydroxide has also been examined in the presence and absence of phosphonate additives using the seeded technique [15]. This salt has wide potential use in water treatment processes as a recyclable coagulant. Recently, Black, et al. [4-6] proposed a water softening and clarifying procedure using $Mg(OH)_2$ as the coagulant that can be recovered and reused. Typical concentration versus time curves are shown in Figure 6.16. The growth experiments with a ratio $T_{Mg}/T_{OH} = 0.5$ are summarized in Table 6.6. The curves in

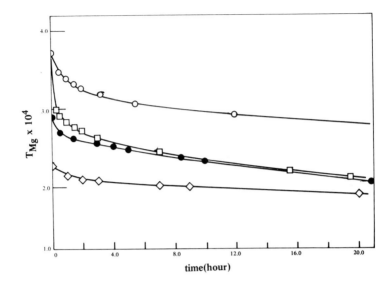

Figure 6.16. Plots of total magnesium concentration as a function of time for magnesium hydroxide crystal growth. ● Expt. 5; □ Expt. 8; ○ Expt. 9; ◊ Expt. 13. Reproduced with permission of *Desalination* [15].

Figure 6.16 are characterized by an initial fast period followed by a smooth fall in concentration with time, and the rate of growth may be written in terms of the changes in total magnesium concentration:

$$-\frac{dT_{Mg}}{dt} = k_c s(T_{Mg} - T_{Mg}^\circ) \qquad (8)$$

Immediately following an initial surge, the plots of the integrated form of Equation 3 become linear, as illustrated in Figure 6.17.

It was found that the duration of the initial surge increases as the degree of supersaturation is increased. This evidence points to additional nucleation during the surge [1], and the relatively slow subsequent process probably includes an aging or recrystallization reaction in addition to the growth of the seed crystals. For this reason, it is seen in Table 6.6 that although there is a

Table 6.6
Crystallization of Magnesium Hydroxide at 25°C, T_{Mg}/T_{OH} = 0.5

Expt. No.	$T_{Mg} \times 10^4$	Seed Conc. mg/100 ml	Additive [A] $\times 10^6$ M		$k'_c s \times 10^2$ h^{-1}
1	3.100	13.8		0	5.9
2	3.078	13.7		0	6.0
3	3.045	13.6		0	5.3
4	3.111	8.3		0	4.0
*5	3.080	8.2		0	4.1
6	3.893	8.2		0	3.6
*7	3.815	10.1		0	3.5
8	3.798	17.0		0	4.6
9	3.858	3.4		0	2.0
10	3.860	3.5		0	2.2
11	2.827	10.1		0	3.8
12	2.244	8.2		0	3.1
13	2.307	8.5		0	3.8
14	2.254	8.1		0	3.0
19	2.836	10.1	NTMP	3.737	2.4
20	2.859	10.2	NTMP	1.868	2.5
21	2.879	10.3	ENTMP	1.862	2.5

*Experiments 5 and 7 (stirring rate reduced to 150 rpm, normal stirring rate at about 350 rpm).

general increase in $k_c s$ with increasing amounts of seed crystals, the increase is not in direct proportion to the latter.

The results of the crystal growth experiment with magnesium hydroxide can be interpreted in terms of a surface diffusion model for crystal growth. According to dislocation theory, both liquid phase and surface diffusion models would predict a rate law linear in relative supersaturation at high supersaturations and a parabolic dependence at very small supersaturations. Since both the liquid phase diffusion and surface diffusion models give analogous results, additional criteria such as the effect of stirring rate changes upon crystal growth must be considered. In Table 6.7 it is seen that a change in the stirring rate of the conductance cell had a negligible effect on the growth rate of magnesium

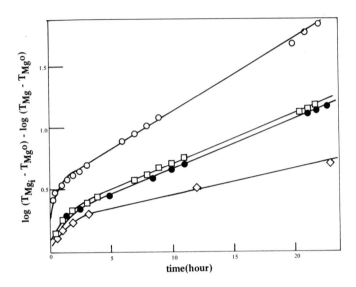

Figure 6.17. Plots of the integrated form of equation 7. ○ Expt. 1; □ Expt. 4; ● Expt. 5; ◊ Expt. 9. Reproduced with permission of *Desalination* [15].

Table 6.7

Crystallization of Magnesium Hydroxide at 25°C, $T_{Mg}/T_{OH} \neq 0.5$

Expt. No.	$T_{Mg} \times 10^4$	$T_{OH} \times 10^4$	Seed Conc. mg/100 ml	$k'_c s \times 10^2$ h^{-1}
15	1.969	7.876	10.5	5.4
16	1.944	7.776	10.4	5.4
17	5.782	5.782	10.3	3.9
18	5.782	5.782	10.1	3.8

hydroxide. This evidence rules out liquid phase diffusion as the rate-determining step for crystal growth. It is interesting to note that a first order kinetic equation similar to Equation 3 satisfactorily represented the rate of crystal growth of both strontium oxalate monohydrate [16] and cadmium oxalate trihydrate [17]. The results are quite different from those for the majority of sparingly soluble salts such as silver chloride [18], strontium sulfate [19] and calcium sulfate dihydrate [20]. In such cases, the rate of crystal growth was found to be proportional to the square of the relative supersaturation.

The results of crystal growth experiments in solutions containing nonequivalent concentrations of magnesium and hydroxide ions are summarized in Table 6.7. The data again follow Equation 8 and it is seen that, when the anion is in excess, k_cs values are larger than for those experiments made in the presence of excess magnesium ions. In an electrophoresis study, Larson and Buswell [21] have shown that precipitated magnesium hydroxide particles carry a positive charge throughout the pH range of interest in the present work. In the supersaturated solutions, therefore, magnesium ions are more strongly adsorbed by the crystal surface than hydroxide ions and when the latter are in excess, the rate constant for crystal growth is increased.

Table 6.6 summarizes some experiments made in the presence of phosphonate additives, which are well-known crystal growth inhibitors. Under such conditions, Equation 8 again satisfactorily interprets the experimental results, and it is seen that the crystallization rate constant is reduced in the presence of both inhibiting additives nitrilotri(methylene phosphonic acid) or NTMP and N,N,N',N'-ethylenediaminetetra(methylene phosphonic acid) or ENTMP. The retarding effect is, however, very much less than in the case of calcium sulfate [20] and calcium carbonate [10] for which the rate of crystal growth follows the square of the supersaturation.

The growth of magnesium hydroxide crystals differs from calcium carbonate in two significant ways. For magnesium hydroxide the rate follows a first order kinetic expression, whereas for calcium carbonate n = 2 in Equation 4. The crystallization of magnesium hydroxide is little affected by the presence of phosphonate ions in solution, whereas for calcium carbonate there is a marked rate reduction. Phosphonate ions adsorb on both magnesium hydroxide and calcium carbonate surfaces. The small growth inhibition in the case of magnesium hydroxide is probably caused either by a relatively weak adsorption onto the positively charged surface [21] or the formation of a highly mobile adsorbed phosphonate species. Clearly, magnesium hydroxide precipitation in water treatment processes will not be as prone to inhibition as calcium carbonate precipitation by low phosphonate concentrations in solution.

Dicalcium Phosphate Studies

The crystal growth of dicalcium phosphate dihydrate (DCPD) from metastable supersaturated solution was studied at a constant pH of 5.6 [22]. Growth was monitored by following the decrease in lattice ion concentrations in solution and determining the amount of base required to maintain the solution pH at a constant value. Plots of total calcium and total phosphorus ion concentrations for a typical dicalcium phosphate dihydrate seeded growth experiment are shown in Figures 6.18 and 6.19.

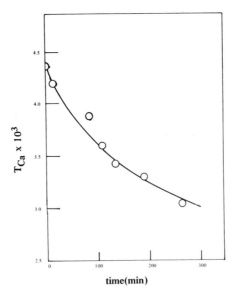

Figure 6.18. Plot of total calcium concentration against time for dicalcium phosphate dihydrate crystal growth, Expt. 37 [22].

Crystal Growth Kinetics of Minerals 245

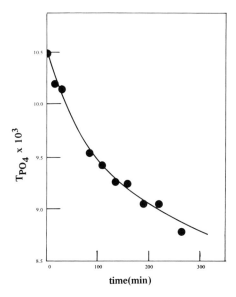

Figure 6.19. Plot of total phosphate concentration as a function of time for dicalcium phosphate dihydrate crystal growth, Expt. 37 [22].

The rate of crystallization of dicalcium phosphate dihydrate follows an equation of the form:

$$- dN/dt = k\, s\, N^2 \qquad (9)$$

where N is the number of moles of salt to be precipitated before equilibrium is reached, and k and s are the rate constant and growth site terms discussed previously. The integrated form of Equation 9 is plotted in Figure 6.20 for the data presented in Figures 6.18 and 6.19. Figure 6.20 illustrates the excellent agreement between the experimental data (Run 37) and the proposed rate law (Equation 9). Several experiments conducted at different stirring rates demonstrate that the rate of crystal growth is independent of solution hydrodynamics. In addition, the activation energy for DCPD growth was found to be 10.5 kcal/mol [23].

Growth of DCPD was monitored in solutions containing phosphonate ions. The experimental conditions are summarized in Table 6.8 [22]. Experiment 38 was performed at a concentration

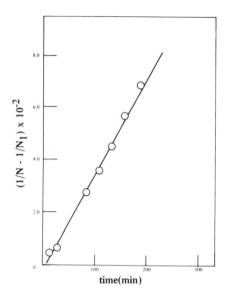

Figure 6.20. Plot of $1/N - 1/N_1$ against time for dicalcium phosphate dihydrate crystal growth, Expt. 37 [22].

of 1×10^{-6} M NTMP which completely inhibited the growth of DCPD. Even at a concentration of 2.5×10^{-7} M additive there was no observable growth for over 24 hours with all four phosphonates. Figure 6.21 shows the effect of 5×10^{-8} M additive for experiments 44, 47, 48 and 46. The order of effectiveness in inhibiting crystal growth appears to be ENTMP > TENTMP > NTMP > EHDP.

Crystal growth experiments were also made in the presence of magnesium ion, and these are summarized in Table 6.8. In Figure 6.22 the amount of base required to maintain a constant pH of 5.6 is plotted as a function of time for a series of magnesium ion experiments. Comparison of Experiments 2A and 5A show that the concentration of magnesium ion additive has to be raised almost to the level of total calcium before significant inhibition is observed. The small effect of 1×10^{-4} M magnesium is shown in Figure 6.23 as the rate plot for Experiment 2A. The significant decrease in the rate of crystal growth for Experiment 4A is shown by the plot of total calcium and total phosphate as

Table 6.8

Crystallization of Dicalcium Phosphate Dihydrate in the Presence of Phosphonate Ions

Expt. No.	$T_{Ca} \times 10^3$ M	$T_{PO_4} \times 10^3$ M	Additive	Additive Concentration M
37	4.37	10.40	—	None
38	4.26	10.35	NTMP	1×10^{-6}
39	4.32	10.30	NTMP	2.5×10^{-7}
44	4.30	10.30	NTMP	5×10^{-8}
43	4.30	10.35	EHDP	2.5×10^{-7}
47	4.40	10.40	EHDP	5×10^{-8}
41	4.38	10.35	ENTMP	2.5×10^{-7}
48	4.30	10.30	ENTMP	5×10^{-8}
42	4.40	10.30	TENTMP	2.5×10^{-7}
46	4.30	10.30	TENTMP	5×10^{-8}
2A	5.60	10.38	—	—
3A	5.52	10.38	Mg^{2+}	1×10^{-4}
4A	5.54	10.33	Mg^{2+}	5×10^{-3}
5A	5.51	10.60	Mg^{2+}	1×10^{-3}
6A	5.30	10.25	Mg^{2+}	7.5×10^{-3}

a function of time in Figure 6.24. The inhibiting effects of magnesium ion may be due to a decrease in relative supersaturation.

Dicalcium phosphate dihydrate crystal growth follows a second order rate law as does calcium carbonate. However, crystal growth rates for the former salt were more markedly reduced by the presence of phosphonate ions in solution. The greater inhibitory effect of phosphonate ions on DCPD growth rates would make it less practical for water treatment programs to employ this salt for the removal of solution phosphate ions in the presence of phosphonates.

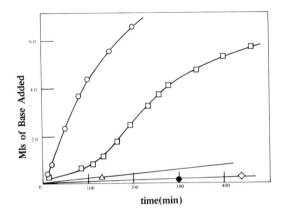

Figure 6.21. Plots of base uptake as a function of time for dicalcium phosphate dihydrate crystal growth at pH 5.6. ◊ Expt. 46; ● Expt. 48; △ Expt. 44; □ Expt. 47; and ○ Expt. 37 [22].

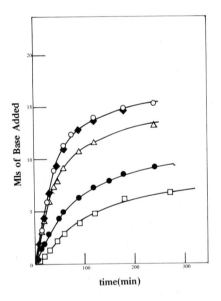

Figure 6.22. Plots of base uptake against time for dicalcium phosphate dihydrate crystal growth at pH 5.6. □ Expt. 6A; ● Expt. 4A; △ Expt. 5A; ♦ Expt. 3A; ○ Expt. 2A [22].

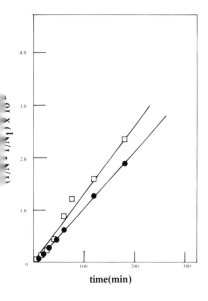

Figure 6.23. Plots of $1/N - 1/N_1$ against time for dicalcium phosphate dihydrate crystal growth in the presence and absence of magnesium ion. ● Expt. 3A; □ Expt. 2A [22].

Figure 6.24. Plots of total calcium and total phosphate concentrations as a function of time for dicalcium phosphate dihydrate crystal growth.
●○ Expt. 4A; ■□ Expt. 2A.

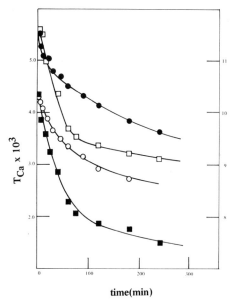

CONCLUSION

The results obtained in this study have allowed formulation of kinetic rate equations for the growth of calcium carbonate, magnesium hydroxide and dicalcium phosphate dihydrate from aqueous supersaturated solutions. These rate equations are applicable to the analysis of crystal growth reactions in water treatment processes. The experimental data for growth of each of the salts point to a rate-determining step for growth involving a reaction at the crystal surface; the rate is not dependent upon diffusion processes in solution.

Crystal growth rates for calcium carbonate and dicalcium phosphate dihydrate have been found to be dramatically reduced in solutions containing phosphonate ions while growth rates of magnesium hydroxide are only slightly reduced in similar solutions. The differences in growth inhibition by phosphonate ions may reflect the extent of adsorption or the formation of a mobile adsorbed phosphonate species on the magnesium hydroxide surface.

These kinetic expressions can be used to calculate the solution contact times required to bring a supersaturated solution to a specified lower concentration level. From this calculation an estimate is obtained of the optimum seed crystal concentration and initial solution supersaturation in a seeded water treatment process. The principal obstacles to the calculation of these parameters in actual water treatment processes are the presence of contaminants such as the polyphosphates and the other potential crystal growth inhibitors in the process feed waters. This has been demonstrated in the growth of calcium carbonate and dicalcium phosphate dihydrate where low concentrations of phosphonate ions greatly reduce the crystal growth rate. In actual plant applications seeded growth measurements can be carried out directly on process feed waters. Comparison with the rate constants determined from work with highly purified reagents and solutions can then be used to estimate the level of crystal growth inhibitors present in the process feed waters.

ACKNOWLEDGMENT

The authors wish to express their appreciation to the Office of Saline Water for financial support through Grant No. 14-30-2633, to Dr. L. Johnson of the University of Western Ontario, London, Ontario, Canada, for the use of the scanning electron microscopy facility in the School of Dentistry at the University of Western Ontario, and to Dr. T. M. King of the Monsanto Company for donating the phosphonic acid derivatives used in this study. The magnesium hydroxide and calcium phosphate studies were done by Dr. S. T. Liu and Mr. J. Wefel, respectively.

NOTATIONS

A	- seed crystal specific surface area, m^2/g
[A]	- molarity of ion A
$[A]_o$	- equilibrium molarity of ion A
$[A]_i$	- initial molarity of ion A
f_z	- activity coefficient (molar scale) of ion of charge z
I	- ionic strength (molar)
K_{sp}	- thermodynamic solubility product
k	- rate constant for crystal growth
k_o	- rate constant for crystal growth in the absence of additives
k_1	- desorption rate constant for ENTMP adsorbed on a calcite surface
k_2	- rate constant for ENTMP adsorption onto a calcite surface
m	- solution concentration
m_o	- saturated solution concentration
N	- number of moles of salt to be precipitated from solution before equilibrium is reached
N_1	- number of moles of salt to be precipitated from the initial solution before equilibrium is reached
S	- solution supersaturation
s	- crystal surface area available for growth, a function of the number of effective crystal growth sites
t	- time
T_A	- total molarity of ion A in solution
T_A^o	- total equilibrium molarity of ion A in solution
T_A^i	- total initial molarity of ion A in solution
α	- fraction of crystal growth sites on a calcite surface covered by adsorbed ENTMP ions

NTPP	- sodium tripolyphosphate
ENTMP	- N,N,N',N'-ethylenediaminetetra(methylenephosphonic) acid
TENTMP	- N,N,N',N'-triethylenediaminetetra(methylenephosphonic) acid
NTMP	- nitrilotri(methylenephosphonic) acid
EHDP	- ethane-1-hydroxy-1,1-diphosphonic acid
DCPD	- dicalcium phosphate dihydrate

REFERENCES

1. Stumm, W. and J. J. Morgan. *Aquatic Chemistry.* (New York: John Wiley and Sons, 1971).
2. Morgan, J. J. in *Equilibrium Concepts in Natural Water Systems.* (Washington, D.C.: American Chemical Society, 1967).
3. Judkins, J. F. and R. H. Wynne, Jr. *J. Amer. Water Works Assoc., 64,* 306 (1972).
4. Black, A. P., B. S. Shisey and P. J. Fleming. *J. Amer. Water Works Assoc., 63,* 616 (1971).
5. Thompson, C. G., J. E. Singley and A. P. Black. *J. Amer. Water Works Assoc., 64,* 11 (1972).
6. Thompson, C. G., J. E. Singley and A. P. Black. *J. Amer. Water Works Assoc., 64,* 93 (1972).
7. Nancollas, G. H. *Interactions in Electrolyte Solutions.* (Amsterdam: Elsvier, 1966).
8. Davies, C. W. *Ion Association.* (Washington, D.C.: Butterworths, 1963), p. 38.
9. Nancollas, G. N. and N. Purdie. *Quart. Rev. (London), 18,* 1 (1964).
10. Reddy, M. M. and G. H. Nancollas. *Desalination, 12,* 61 (1973).
11. Reddy, M. M. and G. H. Nancollas. *J. Colloid Interf. Sci., 36,* 166 (1971).
12. Nancollas, G. H. and M. M. Reddy. *J. Colloid Interf. Sci., 37,* 824 (1971).
13. Nancollas, G. H. and M. M. Reddy. Unpublished data.
14. Raistrick, B. *Discuss. Faraday Soc., 5,* 234 (1949).
15. Liu, T. S. and G. H. Nancollas. *Desalination, 12,* 75 (1973).
16. Gardner, G. L. and G. H. Nancollas. Unpublished data.
17. Liu, T. S. and G. H. Nancollas. Unpublished data.
18. Davies, C. W. and G. H. Nancollas. *Trans. Faraday Soc., 51,* 818, 823 (1955).
19. Campbell, J. R. and G. H. Nancollas. *J. Phys. Chem., 73,* 1735 (1969).
20. Liu, S. T. and G. H. Nancollas. *J. Crystal Growth, 6,* 281 (1970).
21. Larson, T. E. and A. M. Buswell. *Ind. Eng. Chem., 32,* 132 (1940).
22. Wefel, J. W. Ph.D. Thesis. The State University of New York at Buffalo, Buffalo, N.Y. (1974).

23. Marshall, R. W. and G. H. Nancollas. *J. Phys. Chem.*, *73*, 3838 (1969).
24. Greenwald, I. *J. Biol. Chem.*, *141*, 789 (1941).
25. Garrels, R. M. and M. E. Thompson. *Am. J. Sci.*, *260*, 57 (1962).
26. Chughtai, A., R. Marshall and G. H. Nancollas. *J. Phys. Chem.*, *72*, 208 (1968).
27. Greenwald, I., et al. *J. Biol. Sci.*, *135*, 65 (1940).
28. Langmuir, D. *Geochim. Cosmochim. Acta*, *32*, 835 (1968).

CHAPTER 7

EQUILIBRIUM MODELS AND PRECIPITATION REACTIONS FOR CADMIUM(II)

Walter J. Weber, Jr.
Department of Civil Engineering
The University of Michigan
Ann Arbor, Michigan

Hans S. Posselt
Environmental Control Technology
Monroe County Community College
Monroe, Michigan

Introduction 255
Complexation and Distribution Diagrams 256
Solubility Equilibria 261
  Chloride and Ammonia Complexation 264
  Model Validity 269
Precipitation Kinetics 270
  Methods and Materials 272
  Precipitation Experiments 274
  Treatment of Data 275
Results and Discussion 277
  Effect of Carbonate 278
  Effect of pH 280
  Ionic Strength and Anion Effects 283
  Effect of Seeding 285
Summary and Conclusions 287

INTRODUCTION

The effects of trace amounts of cadmium on human health are well-documented [1-4]. In contrast to its relatively more abundant sister element zinc, cadmium is considered nonessential

to man, and its presence in human tissue is regarded as abnormal
[5]. There is sufficient evidence in the literature to conclude
that cadmium accumulates in biological tissues [6], and that it
competes with the essential element zinc for building sites within
certain protein complexes [7]. In view of its potential toxicity
and long biological halflife, it has been recommended that releases of cadmium to the environment be controlled as much
as possible [2].

A high percentage of the cadmium present in the fresh water
and atmospheric environments stems from industrial activities.
For example, in 1968, 35% of the total U.S. consumption of 13.3
million pounds of the metal was eventually emitted to the atmosphere [8]. The technology of removing cadmium and related
metals from industrial wastes and potable water supplies has received limited study in the past, and the conventional approach
of precipitation at elevated pH [9,10] may in many instances be
inadequate in view of more stringent effluent and drinking water
standards [11].

This paper discusses pertinent aspects of the equilibrium
chemistry of cadmium in aqueous solutions and of the kinetics
of precipitation, with particular reference to the treatment of
industrial wastes.

COMPLEXATION AND DISTRIBUTION DIAGRAMS

Cadmium in aqueous solution has a pronounced tendency
to form soluble complexes with both organic and inorganic
ligands. Within the framework of this investigation, only inorganic ligand interactions are considered. Of particular interest
to natural water and industrial wastewater systems are complexes
formed by combination with hydroxide, chloride, and to some
extent ammonia.

The equilibria pertaining to the hydroxy complexes of
cadmium can be represented in terms of the stepwise formation
of these species:

$$Cd^{2+} + OH^- \rightleftharpoons CdOH^+; \quad K_1 \qquad (1)$$

$$CdOH^+ + OH^- \rightleftharpoons Cd(OH)_2; \quad K_2 \qquad (2)$$

Precipitation Reactions for Cadmium(II) 257

$$Cd(OH)_2 + OH^- \rightleftharpoons HCdO_2^- + H_2O; \quad K_3 \quad (3)$$

$$HCdO_2^- + OH^- \rightleftharpoons CdO_2^{2-} + H_2O; \quad K_4 \quad (4)$$

Values for the stepwise formation constants, K_1 through K_4, are given in Table 7.1. For reasons of convenience it is customary to replace the stepwise formation reactions by overall formation reactions. The latter are mathematical representations only and do not depict elementary reaction steps. Stepwise formation constants $K_1 \ldots K_4$ and overall formation constants $\beta_1 \ldots \beta_4$ are related by

$$\beta_1 = K_1$$

$$\beta_2 = K_1 \cdot K_2$$

$$\beta_3 = K_1 \cdot K_2 \cdot K_3$$

$$\beta_4 = K_1 \cdot K_2 \cdot K_3 \cdot K_4$$

$$Cd^{2+} + OH^- \rightleftharpoons CdOH^+ \quad ; \quad \beta_1 \quad (5)$$

$$Cd^{2+} + 2OH^- \rightleftharpoons Cd(OH)_2 \quad ; \quad \beta_2 \quad (6)$$

$$Cd^{2+} + 3OH^- \rightleftharpoons HCdO_2^- + H_2O \quad ; \quad \beta_3 \quad (7)$$

$$Cd^{2+} + 4OH^- \rightleftharpoons CdO_2^{2-} + 2H_2O \quad ; \quad \beta_4 \quad (8)$$

A mass balance on soluble cadmium species with a total of C moles/l as Cd, therefore, is given by:

$$C = [Cd^{2+}] + [CdOH^+] + [Cd(OH)_2] + [HCdO_2^-] + [CdO_2^{2-}] \quad (9)$$

Substituting in terms of the formation constants from Equations 5 through 8 gives

$$C = [Cd^{2+}](1 + \beta_1[OH^-] + \beta_2[OH^-]^2 + \beta_3[OH^-]^3 + \beta_4[OH^-]^4) \quad (10)$$

If α is defined as the fraction of a particular species relative to the total analytic concentration, then $[A] = \alpha_A \cdot C$ and the sum

Table 7.1
Equilibrium Constants Used for Calculation of Cadmium Distribution and Solubility Diagrams

$Log_{(10)}$ of Equilibrium Constants

Ligand	K_1	K_2	K_3	K_4	K_5	K_6	Conditions	Reference
OH^-	4.16	4.23	0.69	-0.32	—	—	zero ionic strength, 25°C	[13]
Cl^-	1.32	0.90	0.09	-0.45	—	—	4.5 M $NaClO_4$, 25°C	[14]
NH_3	2.65	2.10	1.44	0.93	-0.32	-1.66	2 M NH_4NO_3, 25°C	[13]

Overall Formation Constants

Ligand	β_1	β_2	β_3	β_4	β_5	β_6	Conditions	Reference
OH^-	$1.45 \cdot 10^4$	$2.46 \cdot 10^8$	$1.20 \cdot 10^9$	$5.76 \cdot 10^8$	—	—	zero ionic strength, 25°C	[13]
Cl^-	21.0	166.0	204.0	71.5	—	—	4.5 M $NaClO_4$, 25°C	[14]
NH_3	$4.47 \cdot 10^2$	$5.63 \cdot 10^4$	$1.55 \cdot 10^6$	$1.32 \cdot 10^7$	$6.31 \cdot 10^6$	$1.38 \cdot 10^5$	2 M NH_4NO_3, 25°C	[13]

of α's over all species is 1:

$$\sum_{i=0}^{n} \alpha_i = \alpha_0 + \alpha_1 + \ldots + \alpha_n = 1 \tag{11}$$

The fractions of the individual solute species of cadmium then may be calculated as a function of hydroxide concentration. Thus, α_n is independent of the cadmium concentration in solution

$$\alpha_0 = \frac{[Cd^{2+}]}{C} = \frac{1}{1 + \beta_1[OH^-] + \beta_2[OH^-]^2 + \beta_3[OH^-]^3 + \beta_4[OH^-]^4} \tag{12}$$

$$\alpha_1 = \frac{[CdOH^+]}{C} = \beta_1[OH^-]\alpha_0 \tag{13}$$

$$\alpha_2 = \frac{[Cd(OH)_2]}{C} = \beta_2[OH^-]^2\alpha_0 \tag{14}$$

$$\alpha_3 = \frac{[HCdO_2^-]}{C} = \beta_3[OH^-]^3\alpha_0 \tag{15}$$

$$\alpha_4 = \frac{[CdO_2^{2-}]}{C} = \beta_4[OH^-]^4\alpha_0 \tag{16}$$

The distribution of these species as a function of pH is shown in Figure 7.1. The distribution diagram for cadmium hydroxy complexes reveals that the species Cd^{2+}, $CdOH^+$, $Cd(OH)_2(aq)$ and $HCdO_2^-$ contribute increasingly to the solubility of cadmium in the range pH 7-12. The distributions of species for cadmium chloride and amine complexes are presented in Figures 7.2 and 7.3, respectively. In accord with these diagrams, complexation by chloride ion becomes a significant factor in solubility considerations of systems containing greater than 10^{-3} M chloride, whereas the role of ammonia as a ligand appears rather insignificant considering its relatively low concentration in most natural waters and wastewaters.

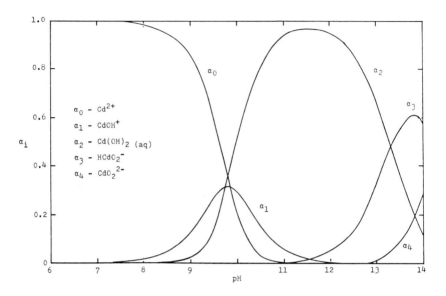

Figure 7.1. Distribution diagram for cadmium hydroxide complexes.

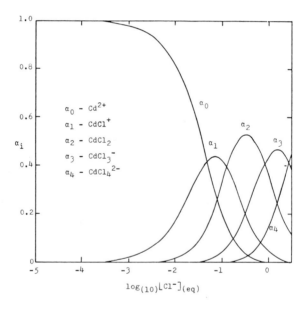

Figure 7.2. Distribution diagram for cadmium chloride complexes.

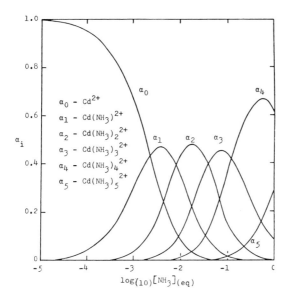

Figure 7.3. Distribution diagram for cadmium amine complexes.

SOLUBILITY EQUILIBRIA

In most environmental circumstances the solubility of cadmium is governed by hydroxide or carbonate. The corresponding solubility product expressions are

$$K_{sp} = [Cd^{2+}][OH^-]^2 = 2.2 \times 10^{-14} \qquad (17)$$

$$K'_{sp} = [Cd^{2+}][CO_3^{2-}] = 5.2 \times 10^{-12} \qquad (18)$$

Solubility calculations based on these solubility products, e.g., $\log[Cd^{2+}]_{(aq)} = 14.34 - 2\,pH$, are valid only to about pH 8, beyond which the contribution of hydroxy-complexes must be taken into account (see also Figure 7.4). In the absence of carbonate in the pH range 7-12, the mass balance for soluble cadmium in terms of formation constants is given by:

$$C = [Cd^{2+}](1 + \beta_1[OH^-] + \beta_2[OH^-]^2 + \beta_3[OH^-]^3) \qquad (19)$$

and substitution of the solubility product, Equation 17, yields an expression for C which is independent of the cadmium concentration as long as the system remains saturated with respect to solid cadmium hydroxide. Under this condition, C is equivalent to the solubility (designated as S)

$$S = \frac{K_{sp}}{[OH^-]^2} (1 + \beta_1 [OH^-] + \beta_2 [OH^-]^2 + \beta_3 [OH^-]^3) \quad (20)$$

In aqueous systems containing both hydroxide and carbonate, the ratio of the concentrations of these ions and of their respective solubility products will determine which of the corresponding K_{sp} values applies to the particular case. Thermodynamically it is required that the relationship yielding the lower concentration of Cd^{2+} governs. Division of the two solubility products gives the ratio

$$R = \frac{[OH^-]^2}{[CO_3^{2-}]} = 0.00423 \quad (21)$$

which is a suitable parameter for comparison of system conditions. If R > 0.00423, Equation 17 applies and if R < 0.00423, S is calculated in accordance with the carbonate solubility product. If R is just equal to 0.00423, either relationship holds.

The "ion-ratio" method has also been employed as a rule for predicting the order of precipitation of solids in cases where two or more precipitation reactions are competing [17]. Because this rule is based on thermodynamic considerations, it may be overruled by kinetic factors under nonequilibrium conditions. The equilibrium concentration of carbonate is obtained from the mass balance taken over all carbonic species

$$C_T = [CO_2^*]^\dagger + [HCO_3^-] + [CO_3^{2-}] = [CO_3]_{TOT} \quad (22)$$

and from the pertinent equilibria for carbonic acid and water, all at 25°C.

$^\dagger [CO_2^*] = [CO_2(aq)] + [H_2CO_3]$

Precipitation Reactions for Cadmium(II) 263

$$K_{a1} = \frac{[H^+][HCO_3^-]}{[CO_2*]} = 4.4 \times 10^{-7} \qquad (23)$$

$$K_{a2} = \frac{[H^+][CO_3^{2-}]}{[HCO_3^-]} = 5.6 \times 10^{-11} \qquad (24)$$

$$K_w = [H^+][OH^-] = 1.0 \times 10^{-14} \qquad (25)$$

Substitution of these expressions into Equation 22 and rearrangement yields the desired term relating $[CO_3^{2-}]$, pH, and total analytic concentration of carbonic acid species:

$$[CO_3^{2-}] = \frac{C_T}{\frac{K_w^2}{[OH^-]^2 \cdot K_{a1} \cdot K_{a2}} + \frac{K_w}{K_{a2}[OH^-]} + 1} \qquad (26)$$

Thus, if R ⩽ 0.00423, the concentration of soluble cadmium is calculated according to

$$S = \frac{K'_{sp}}{[CO_3^{2-}]}(1+\beta_1[OH^-]+\beta_2[OH^-]^2+\beta_3[OH^-]^3) \qquad (27)$$

in which $[CO_3^{2-}]$ is given by Equation 26.

A logarithmic plot of the solubility of cadmium versus pH for different total analytic concentrations of carbonic species and without correcting for ionic activities is presented in Figure 7.4. The effect of carbonate on the solubility of cadmium is very pronounced, suggesting that removal via carbonate precipitation appears more promising, provided that attainment of equilibrium proceeds at practical rates. Conversely, the equilibrium solubilities of cadmium relative to hydroxide would result in effluents much more concentrated than tolerable with regard to fish toxicities, let alone standards for drinking water.

264 *Aqueous-Environmental Chemistry of Metals*

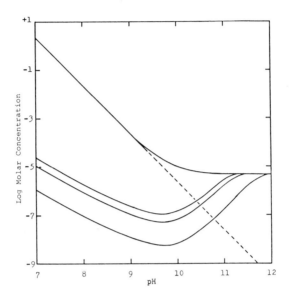

Figure 7.4. Solubility of cadmium as a function of pH at different concentrations of carbonic acid species. The dotted line represents the solubility corresponding to the solubility product of cadmium hydroxide. The solid lines from top to bottom correspond to: $C_T = [CO_2] + [H_2CO_3] + [HCO_3^-] + [CO_3^{2-}] = 0$, 5×10^{-4}, 10^{-3}, and 10^{-2} M, respectively.

Chloride and Ammonia Complexation

In the range of chloride concentration $[Cl^-]_{eq} < 0.1$ M only the species $CdCl^+$ and $CdCl_{2(aq)}$ need be considered. Thus, the mass balance for soluble cadmium in aqueous solution is:

$$S = [Cd^{2+}] + [CdOH^+] + [Cd(OH)_{2(aq)}] + [HCdO_2^-]$$
$$+ [CdCl^+] + [CdCl_{2(aq)}] \tag{28}$$

In the absence of carbonic species and in cadmium hydroxide saturated systems, Equation 28 will take the following form:

Precipitation Reactions for Cadmium(II) 265

$$S = \frac{K_{sp}}{[OH^-]^2} (1+\beta_1[OH^-]+\beta_2[OH^-]^2+\beta_3[OH^-]^3)$$

$$+ \frac{K_{sp}}{[OH^-]^2} (21[Cl^-] + 166[Cl^-]^2) \quad (29)$$

The equilibrium concentration of chloride must be calculated from the mass balance for all chloride species:

$$C_{Cl} = [Cl^-] + [CdCl^+] + 2[Cd(Cl)_{2(aq)}]$$

$$= [Cl^-] + 21[Cd^{2+}][Cl^-] + 2.166[Cd^{2+}][Cl^-]^2$$

$$= [Cl^-] + \frac{21 K_{sp}[Cl^-]}{[OH^-]^2} + \frac{33 K_{sp}[Cl^-]^2}{[OH^-]^2} \quad (30)$$

which involves the solution of a quadratic equation. Similar system equations can be formulated for the case in which the solubility product of cadmium carbonate is governing.

$$S = \frac{K'_{sp}}{[CO_3^{2-}]} (1+\beta_1[OH^-]+\beta_2[OH^-]^2+\beta_3[OH^-]^3)$$

$$+ \frac{K'_{sp}}{[CO_3^{2-}]} (21[Cl^-] + 166[Cl^-]^2) \quad (31)$$

$$C_{Cl} = [Cl^-] + \frac{21 K'_{sp}[Cl^-]}{[CO_3^{2-}]} + \frac{332 K'_{sp}[Cl^-]^2}{[CO_3^{2-}]} \quad (32)$$

The effect of chloride on the solubility of cadmium in aqueous solutions at different pH levels for a carbonate-free system and in 10^{-3} M total analytic carbonate is shown in Figures 7.5 and 7.6, respectively. As evident from these plots, the slopes of the lines in both cases tend to decrease with increasing pH and approach zero as the pH increases above 10.5. This demonstrates, in accord with thermodynamic reasoning, that complexation by chloride can no longer compete with that by hydroxide at sufficiently high pH. The influence of carbonate on solubility is independent of chloride concentration. As is

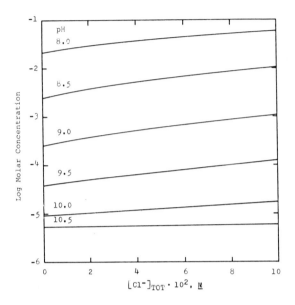

Figure 7.5. Solubility of cadmium as a function of total analytic chloride concentration at different pH values.

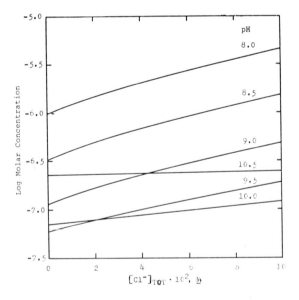

Figure 7.6. Solubility of cadmium as a function of total analytic chloride concentration at different pH values in a system containing 10^{-3} M total carbonic acid species.

Precipitation Reactions for Cadmium(II) 267

true for the calculations shown in Figure 7.4, a tenfold increase of total analytic carbonate yields a tenfold decrease in cadmium concentration. According to Figure 7.3, systems containing ammonia in the range $0\text{-}10^{-2}$ M [NH$_3$] involve the participation of the complex species Cd(NH$_3$)$^{2+}$, Cd(NH$_3$)$_2^{2+}$, and Cd(NH$_3$)$_3^{2+}$ in addition to those formed by interaction with hydroxide. Accordingly, the mass balance for soluble cadmium in the range pH 8-12 and carbonate-free solution, omitting the negligible contribution by the species HCdO$_2^-$, becomes:

$$S = [Cd^{2+}] + [Cd(OH)_{2\,(aq)}] + [CdOH^+] + [Cd(NH_3)^{2+}]$$

$$+ [Cd(NH_3)_2^{2+}] + [Cd(NH_3)_3^{2+}] \qquad (33)$$

and that for ammonia is given by:

$$C_{NH_3} = [NH_3] + [NH_4^+] + [Cd(NH_3)^{2+}] + 2[Cd(NH_3)_2^{2+}]$$

$$+ 3[Cd(NH_3)_3^{2+}] \qquad (34)$$

in which the concentrations of NH$_3$ and NH$_4^+$ as functions of OH$^-$ are related by the equilibrium constant K_b, which has the value 1.78×10^{-5} at 25°C. After a series of familiar substitutions, the two system equations take the form:

$$S = \frac{K_{sp}}{[OH^-]^2} \left[(1+\beta_1[OH^-]+\beta_2[OH^-]^2) + (\beta'_1[NH_3]+\beta'_2[NH_3]^2+\beta'_3[NH_3]^3) \right] \qquad (35)$$

$$C_{NH_3} = [NH_3] + \frac{K_b[NH_3]}{[OH^-]} + \frac{\beta'_1 K_{sp}[NH_3]}{[OH^-]^2} + \frac{2\beta'_2 K_{sp}[NH_3]^2}{[OH^-]^2}$$

$$+ \frac{3\beta'_3 K_{sp}[NH_3]^3}{[OH^-]^2}$$

$$= [NH_3] \left(1 + \frac{K_b}{[OH^-]} + \frac{\beta'_1 K_{sp}}{[OH^-]^2} \right) + [NH_3]^2 \frac{2\beta'_2 K_{sp}}{[OH^-]^2}$$

$$+ [NH_3]^3 \frac{3\beta'_3 K_{sp}}{[OH^-]^2} \qquad (36)$$

Substitution of $K'_{sp}/[CO_3^{2-}]$ for $K_{sp}/[OH^-]^2$ allows calculation of S in carbonate-governed situations.

Consistent with equilibrium data (Table 7.1), complexation of cadmium by ammonia is stronger than by chloride. Since ammonia is rarely present in more than minute amounts in natural waters and usually occurs at levels lower than chloride in industrial wastes, the differences in ligand strengths are ordinarily compensated for by relative concentration ratios. As evident from the slopes of the curves in Figures 7.7 and 7.8, the effect of the total molar concentration of ammonia on cadmium solubility reaches a maximum between pH 8.5 and 10 and levels off at both ends of this pH range. By inspection of appropriate distribution diagrams, this behavior can be directly attributed to formation of ammonium ions in the acid range and increasing predominance of hydroxy-complexes at pH > 10.5.

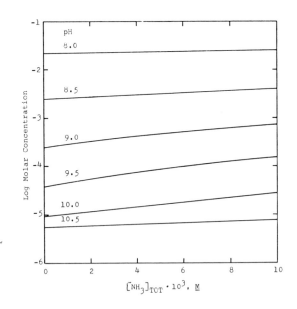

Figure 7.7. Solubility of cadmium as a function of total analytic ammonia concentration at different pH values.

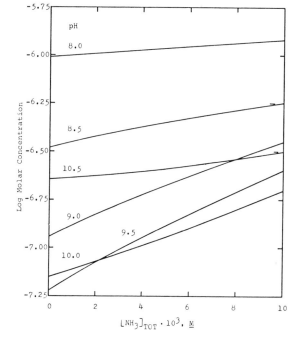

Figure 7.8. Solubility of cadmium as a function of total analytic ammonia concentration at different pH values in a system containing 10^{-3} M total carbonic acid species.

The mathematical treatment of systems containing several ligands simultaneously competing for a central ion becomes increasingly complex. An additional degree of difficulty arises from possible ligand-ligand interactions, as in the case of ammonia and hydroxide. Computation of equilibrium solubilities in multivariable systems may, however, be easily handled with digital computers on the basis of the above outlined mathematical approach [12,15]. In this connection it should be mentioned that noninteracting ligands contribute in an additive fashion to the solubility. Ammonia and chloride complexation of metals illustrates this point.

Model Validity

The models developed for the solubility of cadmium in complex chemical systems are based on the present knowledge of relevant equilibria and of their respective thermodynamic constants. Under the assumptions that the list of species described is exhaustive, that activity corrections are made for individual cases, and that equilibrium constants are numerically factual, the models should predict accurately the solubility of cadmium for a given set of conditions at equilibrium.

There is some uncertainty, however, with regard to the existence of polynuclear hydroxide complexes and of the species $CdHCO_3^+$ and $CdCO_{3(aq)}$. Further, in systems containing multiple ligands there is a possibility for the formation of mixed ligand species. The solubility product for cadmium hydroxide has been found to vary by as much as an order of magnitude, depending on a variety of factors but without any apparent consistency. It is not uncommon to find large discrepancies for solubility products of sparingly soluble salts, depending on the direction of approach to equilibrium. Data measured en route of the dissolution of well-defined crystals may differ substantially from those obtained via precipitation from solution. Phenomena such as phase transitions, aging, colloid formation, and differences in particle size are frequently found to obscure solubility measurements subsequent to precipitation reactions [16,17].

Slow attainment of equilibrium is probably the most significant limiting factor in many applications of thermodynamic solubility relationships, particularly in precipitation reactions.

PRECIPITATION KINETICS

Electroplating, or more generally metal finishing, constitutes perhaps the largest single domestic use of cadmium, primarily for corrosion protection of metal parts [18]. With the exception of chromate the method chosen for treatment of particularly dilute heavy metal-bearing industrial wastes is precipitation under alkaline conditions. Specific treatment considerations for removal of individual metals, including cadmium, are somewhat lacking in the technical literature. In this connection it is worth mentioning that the difference between the solubilities of metal hydroxides and carbonates is often substantial. For example, at pH 9 the solubilities for cadmium are 28.8 and 0.008 ppm (as Cd) for systems containing 0 or 10^{-3} M carbonate, respectively. Because equilibrium is rarely attained under practical conditions, heavy metal levels in industrial effluents may often be much higher than predicted on the basis of thermodynamic calculations. Complexation and colloid stabilization are additional factors in favor of increased effluent concentrations. Dobson [19], for example, reported that effluents from alkaline-chlorinated cyanide wastes contained as much as 60 ppm cadmium.

The primary pollutant in most metal finishing wastes is the extremely toxic cyanide. The most widely accepted method for detoxification of cyanides is alkaline chlorination. Cyanide oxidation with hypochlorite at pH > 10 yields, in the first stage, the much less toxic cyanate

$$CN^- + OCl^- \longrightarrow CNO^- + Cl^- \tag{37}$$

which tends to undergo hydrolysis, resulting in ammonia and carbon dioxide:

$$CNO^- + H_3O^+ \longrightarrow NH_3 + CO_2 \tag{38}$$

Depending on discharge regulations, secondary oxidation of cyanate may be required:

$$2\,CNO^- + 3\,OCl^- + H_2O \longrightarrow N_2 + 2\,HCO_3^- + 3\,Cl^- \tag{39}$$

While the first oxidation step is preferably conducted at high pH to prevent the formation of toxic cyanogen chloride gas,

$$CN^- + HOCl \longrightarrow ClCN + OH^- \tag{40}$$

and hydrogen cyanide, the destruction of cyanate is usually carried out in the range pH 7.5 to 9.

In view of the conditions required for cyanide oxidation, heavy metal precipitation usually coincides directly with the former process but not necessarily with optimal results. In these circumstances carbonate levels and pH may vary significantly, depending on the degree of oxidation and on subsequent chemical operations. Furthermore, the extent of heavy metal removal is strongly influenced by physical treatment conditions, such as rates of mixing and sedimentation times.

Methods and Materials

A tracer technique based on direct Cerenkov counting of aqueous samples was employed in these investigations [20]. An isotope suitable for this purpose is $_{48}Cd^{115m}$ with a halflife of 43 days [21]. It is commercially available (New England Nuclear No. NEZ012) in 99+ per cent radiopurity in the form of $CdCl_2$ in 0.5 N HCl. The decay process for $_{48}Cd^{115m}$ yields $_{49}In^{115}$ by emission of a 1.63 MeV β-particle. Interference by subsequent decay of indium is nonexistent because of its weak decay energy (E_{max} = 0.6 MeV) relative to the requirements for Cerenkov counting ($E_{max} \geqslant$ 1 MeV). A detection efficiency of approximately 35% was realized with this technique by measurement of events in the tritium channel using a Beckman Model CPM-200 scintillation counter. Within the range of 15 to 21 ml, sample volume effects on counting efficiency were found negligible. Vial volumes of 20 ± 1 ml, however, were employed for reasons of consistency.

All samples were acidified (pH < 2) with 4 N HNO_3 prior to radioassay to avoid light absorption due to precipitation and to minimize metal losses by sorption on container walls. Polyethylene vials were found superior to glass for the latter reason. Transfers of radioactive solutions and those containing minute amounts of cadmium were done with labware made of teflon, polypropylene and polyethylene, since these materials were found to exhibit the lowest sorptive affinity towards aqueous species of cadmium. Sample volumes were determined by weight with proper corrections made for differences in specific gravities of individual solution systems. In most cases the statistical counting error was set to ± 2% or better, depending on specific activities. The linearity between tracer concentration and count-rate was established repeatedly and over a wide range of activities.

Membrane filtration was employed for separation of solids from the liquid phase. All components of the filter holder were made of teflon. Sorptive uptake of cadmium by the conventional cellulose base membrane filters (Metricel G A-6, Gelman Instrument Co.) was found relatively small, but became negligible by employment of larger pore sizes (0.45μ), affording faster filtration rates, and by discarding a substantial portion of the initial filtrate.

All chemicals used were of standard reagent quality unless otherwise specified. Stock solutions of 10^{-2} M $Cd(NO_3)_2$ and 10^{-2} M HNO_3 were prepared by dissolution of cadmium metal in 2 M HNO_3 and addition of distilled water. A stock solution of 500 ml of 10^{-4} M of the isotope was made by appropriate dilution of 2 mc $_{48}Cd^{115m}$ with 0.01 M HNO_3 in a polyethylene bottle. All other reagent solutions used in the experiments were prepared and standardized according to *Standard Methods* [22] and other well-established procedures. Frequent titrations of the distilled water used revealed a CO_2 level of 10^{-4} M or less.

A Beckman Zero-matic pH meter with a calomel/glass electrode couple was employed for pH measurements. Precipitation experiments were conducted in 3/8-inch plexiglass reactors (Figure 7.9). The solutions were kept agitated with pyrex-stirrers driven by synchronous motors. Stirring speeds were regulated by adjustable "powerstats" with primary voltage stabilization and measured with a stroboscope. A thermostated waterbath was employed for temperature control. Suction, provided by an

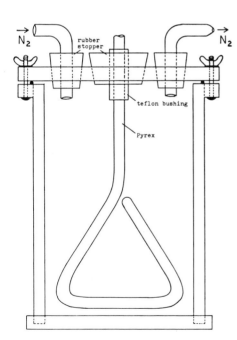

Figure 7.9. Plexiglass-reactor used for precipitation experiments.

aspirator pump, was applied to facilitate filtration of samples. Standard laboratory glassware was used except for handling of cadmium containing solutions, for which devices and containers made of teflon, polypropylene and polyethylene were used.

Precipitation Experiments

Rates of precipitation of cadmium in alkaline carbonate systems were followed by the disappearance of soluble cadmium from solution. All runs were performed in the closed plexiglass reactors at controlled agitator speed, constant temperature and under a blanket of nitrogen. Apart from experimental variations made for evaluation of the influence of one specific variable on the rate of precipitation, the following standard conditions were employed: initial volume 800 ml; temperature 27 ± 0.5°C; agitator speed 400 rpm; ionic strength 0.05 ($NaClO_4$); $[CO_3]_{TOT} = 10^{-2}$ M; $[Cd]_o = 10^{-4}$ M; $[_{48}Cd^{115m}]_o \approx 10^{-6}$ M.

Initially a volume of 780 ml was prepared by mixing predetermined quantities of the reagents, except for $_{48}Cd^{115m}$ and Cd, with distilled water to a given pH. An additional quantity of NaOH was included for neutralization of the acid cadmium solution. After sealing, the reactors were placed in the waterbath and stirred at 400 rpm for approximately 30 minutes under a flow of 0.5 l/min of nitrogen until reaction temperature was reached. Solutions containing either radioactive or stable cadmium were transferred with polyethylene pipettes into a 100-ml teflon beaker and mixed with distilled water to give a total of 20 ml of reactant solutions. At time zero the reactant solution was transferred to the reactor by pouring the total beaker content into an open 30-ml polyethylene syringe (no needle attached) positioned over the N_2-outlet of the vessel. This technique gave reproducible reactant flow conditions and initial precipitation rates. Losses of the reactant by adhesion to walls were found negligible.

Sample portions of approximately 20 ml were withdrawn by means of polyethylene syringes to which a polyethylene tube had been attached. These were directly transferred into the teflon filter funnel and filtered through 0.45μ membranes under slight negative pressure. The first portion of about 20 ml was discarded, whereas the second was directly filtered into the vial attached to the filter assembly. Usually, two identical systems were run in parallel. Thus, filter cleaning or replacement was unnecessary when duplicate samples were taken in succession. Otherwise, the membranes were treated with 1 N HNO_3 and thoroughly washed before each filtration. Soluble cadmium in the filtrate was determined by radioassay as described above.

Treatment of Data

In the range pH 8.5 to 11, $[CO_3]_{TOT} = 10^{-3} - 10^{-2}$ M, ionic strength 0.01 - 0.05 and $[Cd]_o = 10^{-4} - 10^{-3}$ M, the formation of a bulk precipitate was observed to occur almost instantly under moderate mixing conditions. From the standpoint of practical applications, it is of considerable interest to know the extent of initial bulk removal as well as the rate of secondary growth. In view of the complex nature of the systems under study and of the limitations of the experimental technique with respect to studying nucleation and growth in greater detail, it was found appropriate to treat both phases of precipitation on a separate basis. This practice is not uncommon in the study of complex precipitation reactions [23].

Figure 7.10 shows a representative plot for residual cadmium (logarithmic scale) against time for a given set of conditions. The rapid initial fall-off in solute species followed by a steady rate of secondary precipitation or precipitate growth is apparent from this figure and was found typical for all experiments. An empirical rate expression used for description of data for the secondary growth phase is

$$\frac{d\alpha}{dt} = k(1-\alpha) \quad (41)$$

in which α designates the precipitated fraction with values ranging from 0 to 1. By this definition, α in terms of concentrations

$$\alpha = \frac{[Cd]_o - [Cd]_t}{[Cd]_o - [Cd]_s} \quad (42)$$

is dimensionless. C_o and C_t stand for the concentrations of cadmium at time zero and t, respectively, while $[Cd]_s$ is the solubility limited concentration for the respective aqueous system also designated as S. Because $[Cd]_o \gg S$, Equation 42 simplifies to:

$$\alpha = \frac{[Cd]_o - [Cd]_t}{[Cd]_o} \quad (43)$$

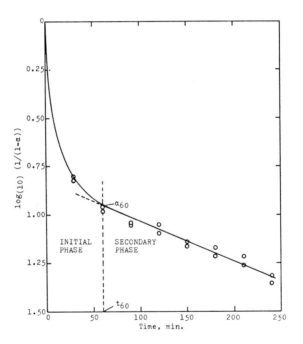

Figure 7.10. Graphical representation of cadmium precipitation data. Experimental Conditions: pH 9.5, T = 27±0.5°C, 400 rpm, $[CO_3]_{TOT}$ = 3.5x10^{-3} M, $[Cd]_o$ = 10^{-4} M, ionic strength = 0.05 (NaClO$_4$).

As shown in the figure, a plot of log $1/(1-\alpha)$ against time yields a linear trace for the secondary growth phase, that is for the time range beyond 60 minutes. A first-order representation was found the most reasonable approximation for fitting of the data over a wide range of experimental conditions.

Precipitation Reactions for Cadmium(II) 277

The demarkation line of 60 minutes was chosen because it represents the minimum time required for accommodation of all precipitation data. While this time boundary between both reaction phases affords a convenient comparison for quantitative treatment of data, it has no mechanistic significance.

First-order rate coefficients, k, were obtained by graphical evaluation of plots of log $1/(1-\alpha)$ against time in which the slope is equal to $k/2.303$. The extent of precipitation at a time of 60 minutes is given in terms of α_{60} values. For convenient calculation of the extent of precipitation under experimental conditions, Equation 41 is integrated between the limits (α_{60}, 60) and (α_t, t), and rearranged

$$\log(1-\alpha_t) = \log(1-\alpha_{60}) - \frac{k(t-60)}{2.303} \quad (44)$$

with the condition that t > 60 minutes.

RESULTS AND DISCUSSION

The validity of the first-order rate relationship for description of the secondary precipitation rate data was verified with three different initial cadmium concentrations. Furthermore, it was confirmed by the linearity of logarithmic rate representations (*e.g.*, Figure 7.10) for a wide range of cadmium concentrations and experimental conditions.

Table 7.2 shows kinetic data for the precipitation of cadmium at different initial cadmium concentrations under the following conditions: pH 9.5; T = 27 ± 0.5°C; 400 rpm; ionic strength 0.05 (NaClO$_4$); [CO$_3$]$_{TOT}$ = 10^{-2} M.

Table 7.2
Effect of Cadmium Concentration on Rate of Precipitation of Cadmium

$[Cd]_o \times 10^5$ M	α_{60}	$\log\left(\dfrac{1}{1-\alpha_{60}}\right)$	k min^{-1}
5.0	0.946	1.27	0.0179
10.0	0.960	1.40	0.0181
20.0	0.980	1.70	0.0182

Most runs were followed to a final cadmium concentration in solution in the range of $\approx 10^{-7}$ M. Thus, from the data presented in the table, it may be concluded that the secondary precipitation rate for cadmium under the conditions studied can be represented by a first-order relationship with respect to cadmium for a range of [Cd] from 10^{-7} to 10^{-4} M.

The initial extent of the reaction, expressed in terms of α_{60}, tends to increase slightly with increasing initial cadmium concentration (Table 7.2) in the range of 5 to 20 x 10^{-4} M. While this observation may, at least in part, be attributed to nucleation phenomena, the constancy of k suggests that the secondary growth rate is fairly independent of the number of nuclei and thus is limited by a different process. In the net-deposition of solute material on nuclei one must also consider the influence of back reactions, that is, the dissolution of preformed crystals which may proceed via various paths, *e.g.*:

$$(A_n B_n)_{solid} \longrightarrow (A_{n-1} B_{n-1})_{solid} + AB_{(aq)}$$

$$(A_n B_n)_{solid} \longrightarrow (A_{n-1} B_n)^-_{solid} + A^+_{(aq)}$$

$$(A_n B_n)_{solid} \longrightarrow (A_n B_{n-1})^+_{solid} + B^-_{(aq)}$$

The rates of these reaction steps are very likely influenced by the solubility of the salt AB and the degree of supersaturation in solution, among other factors [23].

Effect of Carbonate

The concentration of total carbonic acid species, $[CO_3]_{TOT}$, governs the equilibrium solubility of cadmium (*i.e.*, when R < 0.00423) and, as shown in Figure 7.11, is also of strong influence on the overall rate of precipitation. The effect of carbonate concentration on the initial extent of the reaction (α_{60} values) compares well with that observed for cadmium concentration. This is expected as both are "common ions." In view of all experimental evidence, the effect of carbonate on the secondary rate (k values) is interpreted in terms of variations in precipitate composition. At constant pH, the carbonate/hydroxide ratio increases in proportion to the carbonate concentration. Because

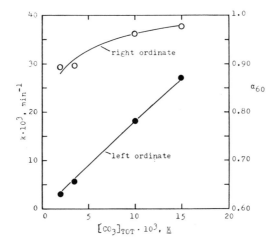

Figure 7.11. Effect of concentration of carbonate on rate of precipitation of cadmium. Experimental Conditions: pH 9.5, T = 27±0.5°C, 400 rpm, $[Cd]_o = 10^{-4}$ M, ionic strength = 0.05 (NaClO$_4$).

the solubilities of both cadmium carbonate and hydroxide are exceeded under these experimental conditions, the composition of initially formed crystals is expected to vary with the ratio of the respective anions. In this light, the secondary growth of CdCO$_3$ crystals at low carbonate levels would be greatly distorted and probably involves the redissolution of Cd(OH)$_2$, while this tendency would decrease in magnitude at higher carbonate/hydroxide ratios.

More definite information on precipitation mechanisms for cadmium carbonate systems would be obtained by study of nucleation and growth kinetics at lower pH, where the solubilities of Cd(OH)$_2$ and CdCO$_3$ are not exceeded concurrently. By the nature of carbonate equilibria, these systems would require controlled carbon dioxide pressure and, consequently, sealing against the atmosphere.

As discussed in the following section, both species, HCO$_3^-$ and CO$_3^{2-}$, can lead to formation of cadmium carbonate; since the rates for carbonate-bicarbonate equilibria are extremely rapid, a kinetic and mechanistic differentiation in terms of cadmium carbonate formation is extremely difficult.

Effect of pH

The variations of solute species and solubility of cadmium as functions of pH suggest a direct and strong influence of this parameter on the kinetics of precipitation of cadmium from alkaline carbonate solutions. Experimental evidence for this is presented in Figure 7.12, which shows the effect of pH on the initial reaction extent, α_{60}, and on the rate of the secondary phase, k, of cadmium precipitation in the presence of 10^{-2} M total carbonate species.

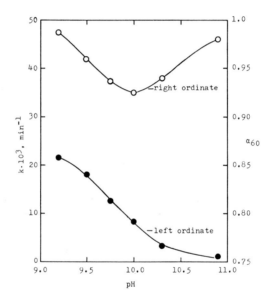

Figure 7.12. Effect of pH on rate of cadmium precipitation. Experimental Conditions: T = 27±0.5°C, 400 rpm, $[CO_3]_{TOT} = 10^{-2}$ M, $[Cd]_o = 10^{-4}$ M, ionic strength = 0.05 ($NaClO_4$).

The equilibrium concentration of CO_3^{2-} increases from approximately 5.3 x 10^{-3} M at pH 9 to about 8.5 x 10^{-3} M at pH 11. The rate of secondary growth decreases over this pH range, however, while the initial extent of the reaction goes through a minimum.

A better understanding of this behavior is gained from inspection of the appropriate distribution diagrams and of possible elementary reactions leading to the reaction products. The following reaction schemes are of interest in this light:

$$Cd^{2+} + HCO_3^- \rightleftharpoons CdHCO_3^+ \qquad (45)$$

$$CdHCO_3^+ + OH^- \rightleftharpoons CdCO_{3(s)} + H_2O \qquad (46)$$

$$CdOH^+ + HCO_3^- \rightleftharpoons CdCO_{3(s)} + H_2O \qquad (47)$$

$$Cd^{2+} + CO_3^{2-} \rightleftharpoons CdCO_{3(s)} \qquad (48)$$

$$CdCO_{3(aq)} \rightleftharpoons CdCO_{3(s)} \qquad (49)$$

$$CdOH^+ + OH^- \rightleftharpoons Cd(OH)_{2(s)} \qquad (50)$$

$$HCdO_2^- + H_2O \rightleftharpoons Cd(OH)_{2(s)} + OH^- \qquad (51)$$

$$Cd(OH)_{2(aq)} \rightleftharpoons Cd(OH)_{2(s)} \qquad (52)$$

Although the relative role and contribution of these reactions to the net-rate of cadmium precipitation is unknown, a relationship between this chemistry and the observed reaction rates as functions of pH seems to exist. Two types of mechanisms are indicated both by the kinetics and by the nature of the products. In the lower pH range, the formation of $CdCO_3$ by reactions of the types shown in Equations 45-49 appears to predominate, whereas the other extreme would be represented by the production of cadmium hydroxide (Equations 50-52) at the high end of the pH scale. In this perspective it must be realized that Cd^{2+} and $CdOH^+$ are the only reasonable species leading to $CdCO_3$, and that their existence ceases beyond pH 12.

The observed rate variation with pH would therefore be consistent with a gradual transition between the formation of cadmium carbonate and hydroxide, whereby the simultaneous and/or transient occurrence of cadmium hydroxide cannot be ruled out for the lower pH range, since the solubility of both reaction products is exceeded under the experimental conditions.

The formation of basic carbonates of the type $(CdCO_3)_x \cdot (Cd(OH)_2)_y$ has been observed under similar reaction conditions [24,25], and frequent inclusions of large amounts of anions in cadmium precipitates seems a common phenomenon [26].

The increase observed for the initial reaction extent beyond pH 10 may in part be explained by rapid formation of cadmium hydroxide (Equation 52) which is superimposed on a very slow precipitation of cadmium carbonate; the latter represents the thermodynamically favored compound. This process is followed by slow crystal growth, with a rate dependent on various factors, including species availability at the solid surface, redissolution of cadmium hydroxide, phase transformation of the crystals and other related phenomena.

From the standpoint of practical applications, both the solubility limit and the rate pattern for precipitation should be considered with respect to cadmium removal. In this perspective, the beneficial effect of carbonate is certainly evident. While bulk removal of cadmium in the presence of sufficient carbonate is relatively fast (α_{60} values in Figure 7.12) between pH 9.2 and 10.9, optimum conditions for secondary growth tend to coincide with the pH range of minimum solubility (Figure 7.4). This conclusion derives from rate data and solubility calculations. For purposes of illustration, the times for 99.9% cadmium removal ($\alpha_{99.9}$) at different pH values have been calculated from the data presented in Figure 7.12 in accord with a modified form of Equation 44:

$$t_{99.9} = \frac{6.909 + 2.3 \log(1-\alpha_{60})}{k} + 60 \qquad (53)$$

These 99.9% removal times are given in Table 7.3.

Table 7.3

Calculated 99.9% Removal Times for Cadmium as a Function of pH

pH	9.2	9.5	9.75	10.0	10.3	10.9
$t_{99.9}$, min	177	264	360	590	1360	2460

Note: Conditions cited under Figure 7.12.

Thus, at pH > 10, conditions for cadmium removal are not favorable as indicated by a rapid increase in the required time ($t_{99.9}$) and by the corresponding solubility. A similar situation exists in the range of pH < 9.3. The solubility increases rapidly, and the net-rate of precipitation is expected to fall off concurrent with the decreasing stability of the species CO_3^{2-} and HCO_3^- when pH is lowered.

Ionic Strength and Anion Effects

Small amounts of soluble impurities are known to alter both the growth rate and crystal habit of precipitates [23]. Growth usually tends to be inhibited by foreign ions, mainly due to distortion of the crystal lattice via adsorption on growth sites. Certain ionic or molecular species may also exert specific effects, such as complexation or participation in the growth of crystals. Also, coprecipitation of several salts, as has been discussed for cadmium carbonate and hydroxide, is frequently observed in multicomponent systems. This effect is especially pronounced when the solubility of all reaction products is exceeded.

Unless otherwise specified, all precipitation experiments were carried out at a constant ionic strength of 0.05, which was adjusted by addition of appropriate amounts of sodium perchlorate. In order to examine specific anion effects, the variation of the rate of precipitation was studied over a range of concentrations of different anions representative of wastewaters and natural waters. As shown by a plot of the initial reaction extent, α_{60}, and that of the secondary growth rate, k, against the perchlorate-ionic strength in Figure 7.13, the anion concentration tends to effect nucleation to a higher degree than it does the secondary growth phase within the investigated range of ionic strength. In view of lattice distortions by minute amounts of foreign ions [23], higher rates of precipitation are expected when such ions are completely excluded from the experimental systems (e.g., $[NaClO_4] = 0$). These conditions are of little practical significance, however.

Other ions of interest, in addition to hydroxide and carbonate, are chloride and sulfate, which are omnipresent in natural water systems although in relatively low concentrations. At a constant

284 *Aqueous-Environmental Chemistry of Metals*

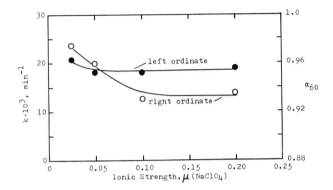

Figure 7.13. Effect of ionic strength on the rate of cadmium precipitation. Experimental Conditions: pH 9.5, T = 27±0.5°C, 400 rpm, $[CO_3]_{TOT} = 10^{-2}$ M, $[Cd]_o = 10^{-4}$ M.

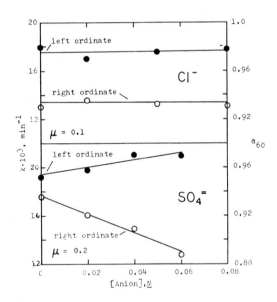

Figure 7.14. Effect of chloride and sulfate on the rate of cadmium precipitation. Experimental Conditions: pH 9.5, T = 27±0.5°C, 400 rpm, $[CO_3]_{TOT} = 10^{-2}$ M, $[Cd]_o = 10^{-4}$ M.

ionic strength of 0.1 as shown in Figure 7.14, chloride ion in the range of 0 to 8 x 10^{-2} M (as NaCl) does not appear to alter the overall rate of cadmium precipitation under the experimental conditions (pH 9.5, $[CO_3]_{TOT}$ = 10^{-2} M, $[Cd]_o$ = 10^{-4} M) investigated. The effect of chloride ion complexation on the solubility of cadmium (Figure 7.6) is negligible in view of the high initial concentration of cadmium, and is thus of no measurable consequence on the rate in terms of α (as defined by Equations 42 and 43). Thus, the anions ClO_4^- and Cl^- exhibit similar effects on the precipitation of cadmium within the range of experimental conditions investigated.

In contrast to monovalent ions, sulfate has a notable effect on the overall rate of cadmium precipitation, as indicated by the data in Figure 7.14. The influence exerted on the secondary growth rate is relatively small and its significance therefore is questionable. The strong dependence of the initial extent of the reaction on the sulfate ion concentration points toward a more pronounced distortion of the crystal lattice by sulfate in the nucleation phase, but alternative explanations cannot be ruled out in the absence of confirmative evidence.

Effect of Seeding

Because flexibility with respect to manipulation of the chemistry and, more particularly, mechanical parameters is often limited in large scale applications of precipitation reactions in water treatment, other approaches to enhancing reaction rates are often more attractive.

In chemical precipitation, the method of "seeding" appears very promising in several respects. It obviates the formation of nuclei with the usual induction period, but, more importantly, it tends to accelerate the rate by propagation of growth of higher ordered and more stable crystals. This implies that negative growth factors, such as dissolution of small crystals and phase transitions, may be minimized or completely eliminated when seed crystals are employed. In highly saturated systems, both the propagation of seed crystals and formation of new particles capable of growth are expected to occur simultaneously. Also, the effect of age and morphology of the seed crystals used must be considered as important rate variables.

With respect to industrial water treatment applications of precipitation processes, recirculation of "sludge" containing preformed crystals would constitute the more practical version of seeding. This method is most feasible in continuously operating systems, but may also be used in batch operations. To examine the feasibility of sludge recirculation, seed crystals were grown under the same pH, ionic strength and $[CO_3]_{TOT}$ conditions as the precipitation experiments. They were prepared approximately 24 hours before the precipitation experiments and were added from well-stirred suspensions prior to the addition of cadmium.

The effect of seeding with previously precipitated cadmium carbonate on the rate of precipitation is shown in Figure 7.15. Both rate components, but most particularly α_{60}, were found to increase considerably with increasing seed concentration. In view of the high initial effect followed by a relatively fast rate of

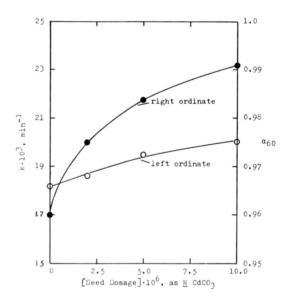

Figure 7.15. Effect of seeding on the rate of cadmium precipitation. Experimental Conditions: pH 9.5, 400 rpm, T = 27±0.5°C, ionic strength = 0.05 (NaClO$_4$), $[CO_3]_{TOT} = 10^{-2}$ M, $[Cd]_o = 10^{-4}$ M.

Precipitation Reactions for Cadmium(II) 287

secondary growth, addition of seed crystals, in both batch or continuous mode, appears to be a very promising and feasible method of attaining high efficiency for cadmium removal by precipitation in industrial wastewaters. In this connection it should be mentioned that addition of precipitated calcium carbonate of up to 10^{-4} M was found to have no measurable effect on the rate of cadmium precipitation under similar conditions.

SUMMARY AND CONCLUSIONS

Models describing the equilibrium solubility of cadmium over a wide range of variable conditions representative for natural waters and wastewaters have been formulated. In accord with these models, the solubility of cadmium in most natural waters should be governed by carbonate or by hydroxide in systems having very low carbonate concentrations. For example, the solubilities of cadmium at pH 8.3 and at concentrations of 0 and 5 x 10^{-4} M total carbonate species are 637 mg/l and 0.11 mg/l, respectively. Simple solubility calculations for cadmium are valid only in the range pH < 9, above which formation of hydroxide complexes becomes increasingly important. Complexation by chloride and ammonia appears insignificant in natural fresh waters in view of the low levels at which these ions occur in such waters, but these reactions may be of serious consequence in the treatment of wastewaters. The latter systems likely will suffer even greater complexity by interaction of cadmium with organic ligands, a recognized tendency of this metal.

Isotope techniques, using $_{48}Cd^{115m}$ and subsequent Cerenkov counting of aqueous samples, are ideally suited for experimental examination of systems containing trace amounts of cadmium. Complications arising from losses of the metal by sorption on container walls can be circumvented or minimized by acidification and/or employment of polyethylene or teflon labware.

The kinetics of precipitation of cadmium in alkaline carbonate solutions has been studied as a function of pH, ionic strength and concentration of cadmium, chloride, sulfate and carbonate species. The rate of precipitation generally exhibits rapid initial and slow secondary growth components; the latter is reasonably well-described by a first-order rate expression with respect to cadmium.

Precipitation of cadmium with carbonate is advantageous in that it results in minimum solubility and faster overall rates as compared with rates and limits of solubility in carbonate-free systems. The variation of rate with pH and with the concentration of carbonate species is indicative of simultaneous operation of several mechanisms leading to mixed reaction products such as basic carbonates. High carbonate/hydroxide ratios represent optimum conditions for quantitative and rapid cadmium removal. These conditions tend to coincide with the pH range 9.3-10.0 of minimum cadmium solubility in carbonate-controlled systems.

The initial extent of the reaction was found to decrease with increasing perchlorate ionic strength and sulfate concentration, while the secondary precipitation rate was not appreciably affected. Chloride exerts no specific effect on the reaction over that observed with perchlorate ion.

Seeding with preformed particulate $CdCO_3$ considerably accelerates the overall rate of subsequent precipitation of cadmium in carbonate systems. Experimental results suggest the practice of sludge recirculation in the treatment of cadmium-bearing wastes. Pilot studies with actual waste solutions should be conducted to evaluate the performance and practical feasibility of this process on a larger scale. Such an approach should also yield further information on the effect of physical conditions (*e.g.*, mixing, settling) and on the general compatibility of cadmium precipitation with a simultaneous removal of other heavy metals.

REFERENCES

1. Friberg, L., M. Piscator and G. Nordberg. *Cadmium in the Environment.* (Cleveland, Ohio: Chemical Rubber Co. Press, 1971).
2. McCaull, J. *Environment, 13,* 3 (1971).
3. Schroeder, H. A. *New England J. Med., 280,* 836 (1969).
4. "Mercury: Other Metals Suspect," *Chem. and Eng. News.* (August 10, 1970), p. 14.
5. Schroeder, H. A. and J. J. Balassa. *J. Chron. Dis., 14,* 236 (1961).
6. "Water Quality Criteria." Report to the National Technical Advisory Committee to the Secretary of the Interior. (Washington, D.C.:

Federal Water Pollution Control Administration, April 1, 1968).
7. Kagi, J. H. and B. L. Vallee. *J. Biol. Chem.*, *236*, 2435 (1961).
8. Malin, H. M., Jr. *Environ. Sci. Technol.*, *9*, 754 (1971).
9. Nemerow, N. L. *Theories and Practices of Industrial Waste Treatment*. (Reading, Mass.: Addison Wesley, 1963).
10. "Methods for Treating Metal Finishing Wastes," (Cincinnati, Ohio: Ohio River Valley Water Sanitation Commission, January 1965).
11. "Inadequate Treatments," *Chem. and Eng. News.* (November 29, 1971), p. 13.
12. Posselt, H. S. Ph.D. Thesis, University of Michigan, Ann Arbor, (1971).
13. Butler, J. N. *Ionic Equilibrium—A Mathematical Approach*. (Reading, Mass.: Addison Wesley, 1964).
14. Bjerrum, J., G. Schwarzenbach, and L. G. Sillen. *Stability Constants*. (London: The Chemical Society, 1957).
15. Morel, F. M. and J. J. Morgan. *Environ. Sci. Technol.*, *6*, 58 (1972).
16. Stumm, W. and J. J. Morgan. *Aquatic Chemistry*. (New York: Wiley-Interscience, 1970).
17. Freiser, H. and Q. Fernando. *Ionic Equilibria in Analytical Chemistry*. (New York: Wiley, 1963).
18. Kirk-Othmer. *Encyclopedia of Chemical Technology*, second ed., Vol. 3. (New York: Wiley, 1964), p. 884.
19. Dobson, J. G. *Sewage Works J.*, *19* (1947).
20. Elrick, R. H. and R. P. Parker. *Intl. J. Applied Radiation Isotopes*, *19*, 263 (1968).
21. Lederer, L. M., J. M. Hollander and I. Perlman. *Table of Isotopes*, 6th ed. (New York: Wiley, 1967).
22. American Public Health Association. *Standard Methods for the Examination of Water and Waste Water*, 12th ed. (New York: American Public Health Association and American Water Works Association,
23. Walton, A. G. *The Formation and Properties of Precipitates*. (New York: Wiley-Interscience, 1967).
24. *Gmelins Handbuch der Anorganischen Chemie*, 8. Auflage, Cadmium Erganzungsband, System Nr. 33. (Weinheim/Bergstr.: Verlag Chemie, 1964).
25. Mellor, J. W. *A Comprehensive Treatise on Inorganic and Theoretical Chemistry*, Vol. IV. (New York: Longmans, Green & Co., 1957).
26. Tananaev, N. V. *Zhur. Neorg. Khim.*, *1*, 2225 (1956).

CHAPTER 8

STUDIES ON THE AQUEOUS CORROSION CHEMISTRY OF CADMIUM

Hans S. Posselt
Environmental Control Technology
Monroe County Community College
Monroe, Michigan

Walter J. Weber, Jr.
Department of Civil Engineering
The University of Michigan
Ann Arbor, Michigan

Introduction 291
Experimental Methods 296
   Electrolytic Deposition of Cadmium 297
   Chemicals and Equipment. 298
   Procedure and Analysis 299
   Treatment of Data 300
Results and Discussion 302
   Cathodic Processes and Mass Transport 302
   Effect of pH 305
   Effect of Chemical Corrosion Inhibitors 306
   Corrosion with Galvanic Coupling 310
Summary and Conclusions 313

INTRODUCTION

Corrosion of zinc-galvanized pipe constitutes a significant source of cadmium in drinking water [1-3]. Although cadmium is a recognized and virtually universal impurity of zinc [4-5], its concentration in the crude grades of this material commonly used for galvanic coating of steel is not ordinarily specified [2].

Copper has gradually become the household plumbing material of choice for new construction, but extensive systems of galvanized steel still exist; it is difficult to estimate the relative proportions of these two types of pipe presently in use. Some studies have indicated that copper pipes may also constitute a source of cadmium [2], but corrosive attack of strictly copper systems probably plays a minor role in the entrance of cadmium to water distribution systems. A more severe situation is likely to exist in cases where galvanized pipes are coupled to copper or brass pipes or fittings because this condition enhances the rate of corrosion of the less noble metals.

The eventual answer to the problem of metal release to drinking waters by corrosion probably lies in the use of suitable plastic plumbing materials [6]. The trend toward plastic pipe is gaining momentum, but reservations have been expressed about certain plasticizers that also may contain toxic heavy metals such as cadmium and lead.

While under more practical conditions both cadmium and zinc are less noble than steel [5,7], cadmium that is contained within the microstructure of a zinc surface and exposed to a corrosive medium is expected to assume the role of the cathode. This implies that under these conditions, only the less noble zinc is subject to corrosive attack. However, it will be shown that "chemical" attack of cadmium may occur as a simultaneous process superimposed on the galvanic corrosion of zinc in a Zn/Cd couple.

Furthermore, it must be realized that water distribution systems often contain metals of higher cathodic potential, such as copper, brass or even steel [7], that are externally coupled with galvanized zinc. These metals exert a driving force on the corrosion of both zinc and cadmium. Coupling of metals with dissimilar electrochemical stability is a very prominent source of corrosive attack, but there are other ways by which potential differences may be established [7,8]. In water distribution systems such situations may also arise from differences in temperature, in concentration of dissolved species, and in flow conditions.

The presumably chemical attack of metals by a corrosive environment is usually regarded as a galvanic process on a microscale. In other words, chemical or physical (*e.g.*, stress-induced) inhomogeneities may lead to a preferential localization of cathodic and anodic sites within the structure of a single metal. Thus, a differentiation between chemical and galvanic corrosion is merely operational, and does not necessarily reflect the nature of the mechanism involved in the corrosive process.

The standard electrode potential for cadmium is -0.400 volts. Thus, cadmium is thermodynamically unstable in nonoxidizing acids and water under a wide range of conditions. In reality, pure cadmium is quite resistant to nonoxidizing acids. This behavior is explained either in terms of its hydrogen overpotential, or mechanistically by a relatively high energy of activation for the discharge of H^+ ions at the metal surface. Thus, from a kinetic point of view, pure cadmium is expected to be stable under the pH conditions of natural waters and in the absence of other driving forces or attacking species. A higher thermodynamic spontaneity for the oxidation of cadmium

$$Cd \rightarrow Cd^{2+} + 2e^- \qquad (E° = -0.4 \text{ V}) \qquad (1)$$

results from the concomitant reduction of oxygen as a cathodic process

$$O_2 + 4H^+ + 4e^- \rightarrow 2H_2O \qquad (E° = +1.23 \text{ V}) \qquad (2)$$

The equivalent cell potential for the oxidation of cadmium via oxygen reduction

$$2Cd + O_2 + 4H^+ \rightarrow 2Cd^{2+} + 2H_2O \qquad (3)$$

is +1.63 volts at pH 0 as compared with +0.4 volts for the process via the reduction of hydrogen.

Again, favorable thermodynamic conditions are inconclusive with regard to actual reaction rates, but owing to relatively lower effects of oxygen overpotentials [8], corrosive attack of certain metals by oxygen usually proceeds at measurable rates [5]. By rearrangement of the Nernst equation for the oxygen attack of cadmium

$$E = E° - \frac{RT}{nF} \ln\left(\frac{[Cd^{2+}]^2}{P_{O_2} \cdot [H^+]^4}\right) \quad (4)$$

and assuming all other variables to remain at standard state conditions, the variation of the cell potential for the overall process as a function of pH is given by

$$E = 1.63 - 0.059 \cdot pH \quad (5)$$

This relationship shows that the driving force for the reaction decreases by 0.059 volts for every unit increase in pH.

Provided that the reaction proceeds at a measurable rate and that the mechanism does not change with pH, the thermodynamic relationship shown in Equation (5) can be used to predict the corrosion rate of cadmium as a function of pH, at least on a qualitative basis. Usually, the situation is more complicated, and factors such as the formation of insoluble films and changes in the activation energy with pH must be taken into account. Kinetic phenomena are in general not readily predictable and must, therefore, be examined by actual experiment. Unfortunately, information on corrosion rates of cadmium in the aqueous environment is almost entirely lacking, and predictions about its kinetic behavior are highly speculative [7].

In addition to pH and oxygen concentration, the driving force for the corrosion of cadmium is also affected by temperature and the concentration of cadmium in solution. Kinetic temperature effects (*e.g.*, Arrhenius-type dependence) are usually found to outweigh thermodynamic factors under the range of practical conditions in aqueous systems. The fact that oxygen solubility decreases strongly with increasing temperature tends to compensate for temperature effects on corrosion rates in the opposite direction.

One very important aspect of general significance to the corrosive properties of water is that of calcium carbonate saturation, which is governed by the following equilibrium:

$$CaCO_{3(s)} + H^+ \leftrightarrow HCO_3^- + Ca^{2+} \quad (6)$$

In quantitative terms, Equation (6) leads to

$$\log[H^+] = \log[HCO_3^-] + \log[Ca^{2+}] - \log K$$

or

$$pH = pHCO_3^- + pCa^{2+} - pK \quad (7)$$

Further, it can be shown that

$$pK = pK_{sp} - pK_2 \quad (8)$$

in which K_{sp} and K_2 are given by

$$K_{sp} = [Ca^{2+}][CO_3^{2-}] \quad (9)$$

and

$$K_2 = \frac{[H^+][CO_3^{2-}]}{[HCO_3^-]} \quad (10)$$

If the equilibrium pH for the reaction shown in Equation (6) is designated as pH_s and the alkalinity

$$Alk = [HCO_3^-] + 2[CO_3^{2-}] + [OH^-] - [H^+] \quad (11)$$

is substituted for the concentration of bicarbonate, the following expression is obtained:

$$pH_s = (pK_2 - pK_{sp}) + pCa^{2+} + p(Alk + [H^+] - \frac{K_w}{[H^+]}) - p(1 + \frac{2K_2}{[H^+]}) \quad (12)$$

The difference between the pH of saturation and the actual pH is a measure of the saturation with respect to calcium carbonate, known as the Langelier Index (L.I):

$$L.I. = pH - pH_s \quad (13)$$

The The Langelier Index is relevant to corrosion in that waters saturated with $CaCO_3$ (L.I. > 0) have a tendency to form scale, whereas waters that are undersaturated (L.I. < 0) are regarded as "aggressive." While the corrosion protection of water pipes by scale formation is very desirable, it has the disadvantage of obstructing the flow and reducing the transfer of heat in boilers.

The practical approach is usually that of a compromise with only a low degree of saturation being favored. A discussion of further rate-controlling factors relative to the corrosion of cadmium is presented in subsequent sections of this paper.

EXPERIMENTAL METHODS

There are numerous methods for studying the rate of corrosion of metals that may differ drastically in character depending, among other factors and considerations, on the purpose of the test and the nature and extent of the process [7].

The objective of this investigation was to evaluate the nature and magnitude of corrosion of cadmium under conditions closely resembling those of drinking water distribution systems and to establish the influence of system variables on the kinetic course of corrosive attack. The requirements of control and simulation of the chemical environment for such a study can usually be met, whereas the task of duplicating the physical conditions of a representative galvanized surface is rather difficult. Furthermore, the relatively low cadmium content in zinc, combined with slow kinetics, would require long term observations and thus not be practical. For these reasons, the study of corrosion of a pure cadmium surface, which can be physically defined at least on a macro-scale, appeared to be a more logical approach. Attempts for a quantitative extrapolation from a pure cadmium surface to more practical conditions can be made, but the validity of such extrapolations must be regarded as tentative in the absence of other experimental evidence. Nevertheless, it seems appropriate to assume that the influence of chemical factors on rates of cadmium corrosion with pure cadmium will be qualitatively similar to those in a cadmium-zinc system, provided appropriate experimental considerations are taken into account.

In view of an established radiochemical technique for trace measurement of cadmium [12], the kinetics of corrosion may be followed easily by counting aliquots of the aqueous solution in contact with a solid cadmium surface of known activity. Since cadmium can be deposited readily from aqueous solutions, the preparation of a pure cadmium surface by plating on an inert metal, such as platinum, is a feasible experimental method.

Electrolytic Deposition of Cadmium

A cadmium "electrode" was prepared by electroplating cadmium on a smooth sheet (0.1 x 10 x 15 mm^3) of platinum metal from perchloric acid solution containing labelled and stable cadmium.

A total amount of 5 x 10^{-4} moles as Cd was originally employed by mixing 2 mc of the tracer solution (5.14 mg $_{48}$Cd^{115m}) and a solution of cadmium perchlorate, prepared by dissolution of high grade Cd O in HClO$_4$, in a tared teflon beaker. The electrolyte was made 1 M in NaClO$_4$ and adjusted to about pH 1.5 by addition of appropriate quantities of sodium perchlorate, perchloric acid and distilled water. The resulting solution ($\approx$ 70 g), containing 5 x 10^{-4} moles of cadmium, was calibrated by weighing and by radioassay as described in Chapter 7. Prior to plating, the platinum surface was etched by treatment with aqua regia and rinsed with distilled water.

Initially, using the apparatus shown in Figure 8.1, about 50% of the total amount of cadmium was deposited by application of a current of approximately 3 mA for several hours, with mixing provided by a magnetic stirrer. This procedure was repeated for 30 minutes at 2 mA prior to each corrosion experiment. The electrode was then removed, rinsed with distilled water, blotted with lens paper and immersed in the corrosion medium. After corrosion testing, the electrode was rinsed and stored in a dry and shielded bottle. Occasionally the cadmium content in the plating bath was depleted as indicated by a decrease in current. In this case, cadmium was stripped from the surface by current reversal and then redeposited according to the above-mentioned procedures.

Hydrogen evolution was not noticed during electrolysis but was found to commence at currents of about 50 mA and more, as was established by preliminary plating experiments with unlabelled cadmium. Also, it was observed that the cadmium surface obtained at low current densities was very smooth compared to dendritic growth when higher currents were applied.

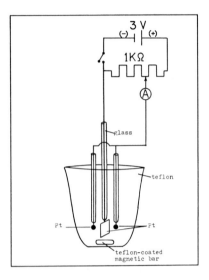

Figure 8.1. Schematic of the assembly used for electrodeposition of cadmium on platinum electrodes.

Chemicals and Equipment

Unless otherwise indicated, all chemicals were reagent grade. A dry and oil-free grade of compressed air and chemically pure nitrogen were used for atmospheric control of the corrosion experiments. When desirable, carbon dioxide was removed by passage of these gases through a column filled with Ascarite, an adsorbent consisting of solid $Ca(OH)_2$, NaOH and certain minor additives. Calgonite, a commercial grade of sodium polymetaphosphate, $(NaPO_3)_x$, was employed in some experiments since this material is used in many water treatment applications including corrosion prevention.

The instruments used for the measurement of pH, activity (scintillation counter) and the speed of agitation have been described in the preceding chapter. A sensitive laboratory-type VOM-meter was used to measure the current during electrolytic deposition of cadmium. Direct current was provided by a heavy load dry-cell battery.

Aqueous Corrosion Chemistry of Cadmium 299

Corrosion experiments were performed in a plexiglass reactor also described previously, with additional openings to accommodate insertion of a thermometer and the cadmium and pH electrodes. The solution was mixed with a 1.5-inch teflon coated stirring bar, which was rotated by a synchronous motor through a magnetic coupling. A medium porosity polyethylene gas immersion tube was used for passage of gases of defined composition through the reaction solution. The desired oxygen content of the atmosphere in contact with the solution was regulated by mixing nitrogen and compressed air. The O_2 concentration in the gas mixture was measured with a polarographic oxygen sensor coupled with a continuous oxygen analyzer, model 77700, both manufactured by Beckman Instruments. Air and O_2-free nitrogen were used as reference gases for calibration of the instrument.

Procedure and Analysis

Prior to addition of chemicals, 770 ml of distilled water contained in the corrosion reactor was degassed by alternate evacuation with an aspirator pump and purging with CO_2-free nitrogen accompanied by rapid mixing. The removal of CO_2 and O_2 was verified by concurrent pH measurement and continuous monitoring of the oxygen concentration in the effluent gas, respectively. Subsequently, and throughout the entire length of the corrosion experiment, the solution was kept saturated with a gas mixture of defined composition. The purging procedure was omitted in the case of experiments in equilibrium with air. In this case air was passed through the solution for 30 minutes prior to corrosion testing.

The desired levels of pH, ionic strength and reagent concentrations were adjusted by addition of appropriate quantities of standardized solutions of the respective chemicals. A total initial volume of 800 ml was prepared for each experiment. Unless otherwise specified, an agitator speed of 450 rpm was maintained during the corrosion experiments. After equilibration of the system, the cadmium "electrode" was immersed in the solution. Its position was fixed so that it would be reproduced in each run.

The rate of corrosion was followed by radioassay of solution aliquots, which were withdrawn from the vessel with a polyethylene syringe and directly transferred into the tared scintillation vials at fixed time intervals. Basically, the same counting technique described previously for the precipitation experiments was employed for radioassay of aqueous samples from the corrosion experiments. Certain modifications became necessary, however, which are described below.

In order to account for the total amount of cadmium in solution as a function of time, the loss of solution during sample withdrawal had to be avoided. Thus, the syringe was rinsed with the sample medium, emptied into the vessel, and after refilling, the total sample volume was transferred into the preweighed vial until the same procedure was repeated. By subtraction of the sample volume as determined by weight the remaining volume in the vessel could be calculated quite precisely.

Since the plating bath volume was bound to fluctuate over the period of this investigation, a different method for standardization was adopted. Prior to any removal of cadmium by deposition, an aliquot of the weighed electrolyte was removed and counted. This standard was used for initial and subsequent calibrations. Repeated checking of the count rate of the initial standard against the decay law for the cadmium isotope revealed excellent agreement, indicating that the instrument efficiency was sufficiently stable. Thus, the concentration of cadmium in solution was calculated from the count rate and the time elapsed from the "time zero" in accordance with the decay law for this isotope. The validity of this procedure was always verified by parallel counting of the initial standard.

Treatment of Data

The rate of corrosion of a uniform metal surface in contact with a large excess of attacking species in the solution phase is expected to be proportional to the surface area of the metal only

$$-\frac{d[Me]}{dt} = kA \qquad (14)$$

Because the surface area, A, is practically constant, the process of metal corrosion under the above specified conditions should approximate a pseudo zero-order reaction

$$-\frac{d[Me]}{dt} = k' \tag{15}$$

This is indeed observed in many corrosion experiments and was found valid also for the corrosion of cadmium in well-stirred aqueous media. Accordingly, a plot of the loss of cadmium per unit area of the metal surface against time should yield a straight line with a slope of k'. Figure 8.2 shows the rate linearity required by zero-order kinetics for the corrosion of cadmium in carbonate-free solution at different pH levels. In accordance with the units used for weight and area, the specific rate constants, k', for cadmium have the units of $\mu g \cdot cm^{-2} \cdot min^{-1}$. These are used

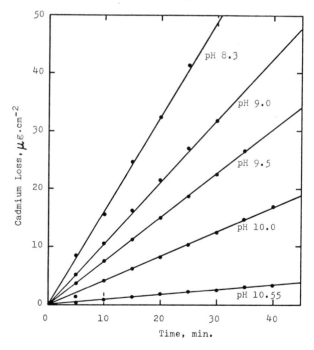

Figure 8.2. Corrosion of cadmium as a function of pH in carbonate-free solutions. Experimental conditions: T = 25.5 ± 0.5°C, Ionic Strength = 0.01 ($NaClO_4$), P_{O_2} = 0.2 atm, 450 rpm.

in all presentations of corrosion rates for cadmium and are determined from the data by the method of least squares with a digital computer.

The proportionality of the rate of corrosion with respect to the surface area of cadmium was established independently by experiments with electrodes of different surface areas. The treatment of data with a nonlinear rate pattern, as was observed in the case of film formation, will be discussed in connection with the interpretation of these individual cases.

RESULTS AND DISCUSSION

Cathodic Processes and Mass Transport

The overall reaction describing the "chemical" dissolution of a single metal in a corrosive aqueous environment is mechanistically divided into cathodic and anodic processes. With the exception of certain relatively noble metals such as copper, anodic reactions of the type

$$Me \rightarrow Me^{n+} + ne^- \qquad (16)$$

are considered very rapid [5]. Thus, the corrosion of cadmium is expected to proceed under cathodic control. Evidence for the assumption that the reduction of oxygen constitutes the cathodic and rate-controlling reaction is shown in Figure 8.3 by a plot of the corrosion rate against the oxygen partial pressure in equilibrium with the solution phase. The rate increases rapidly with increasing concentration of O_2, after which it tends to approach a limiting value. The rate-determining step for the cathodic reaction may be (1) the diffusion of oxygen at the solution-metal interface, (2) a reaction step during ionization of oxygen, or (3) the removal or penetration of diffusion-obstructing solid products, such as $CdCO_3$ or $Cd(OH)_2$ films, or any combination of these phenomena.

Mass transport control is sufficiently evidenced (see also Figure 8.4) for the conditions pertinent to Figure 8.3 and would, by the above definitions, apply to case (1) and (3) equally well.

Aqueous Corrosion Chemistry of Cadmium 303

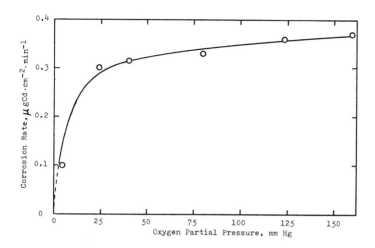

Figure 8.3. Effect of saturation pressure of oxygen on the rate of cadmium corrosion. Experimental conditions: T = 25.5 ± 0.5°C, Ionic Strength = 0.01 (NaClO$_4$), pH 9.5, 450 rpm, [CO$_3$]$_{TOT}$ = 10^{-3} M, P$_{N_2}$ = (760-P$_{O_2}$) mm Hg.

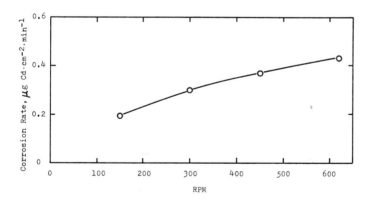

Figure 8.4. Rate of cadmium corrosion as function of agitator speed. Experimental conditions: T = 25.5 ± 0.5°C, Ionic Strength = 0.01 (NaClO$_4$), pH 9.5, P$_{O_2}$ = 0.2 atm, P$_{N_2}$ = 0.8 atm, [CO$_3$]$_{TOT}$ = 10^{-3} M.

Activation control (case 2) can be ruled out on the basis of the data shown in Figure 8.4, because in this situation the rate of corrosion should be independent of the rate of agitation beyond a minimum degree of mixing of the solution. As will be shown in subsequent figures the formation of surface films is indicated by kinetic experiments in carbonate-free and carbonate-containing systems and by the relative rate increase observed with galvanic corrosion experiments performed under comparable conditions.

The observed dependence of the rate on the oxygen concentration is generally consistent with hindrance of oxygen diffusion through a surface film consisting of cadmium carbonate or cadmium hydroxide. In view of an expectedly higher surface concentration of hydroxide ions in the rate limiting region for oxygen (Figure 8.3), the variation of the corrosion rate may also indicate variable film permeability and composition throughout the entire range of oxygen partial pressures investigated.

As shown in Figure 8.4, the rate of cadmium corrosion in well-oxygenated systems increases with an increasing rate of agitation over a fairly wide rpm range without an indication of a plateau region much below approximately 1000 rpm; 450 rpm was chosen for all other experiments since this agitator speed was found most adequate for the desired level of turbulence in the experimental reactor. Although the data indicate that the corrosion rate in quiescent solution is rather slow, it is certainly not expected to become zero in view of convective migration of reactants and products. In fact, convection supported migration in standing pipes may be fairly extensive considering the temperature differences between the incoming water and the environment of its residence.

Cathodic depolarization of cadmium may also be affected by chlorine or hypochlorite ion which are both present in small concentrations, usually less than 1 ppm as Cl_2 in most drinking waters. Corrosion experiments performed with hypochlorite were considered inconclusive because of the observed chemical reduction of chlorine on the plexiglass reactor walls. Because glass reactors are inadequate for reasons of cadmium adsorption a study of the corrosive action of chlorine was not attempted further.

Effect of pH

The corrosion of cadmium was studied in the pH range 8.3 to 10.55 in carbonate-free systems and in the presence of 10^{-3} M total carbonate at an equilibrium oxygen partial pressure of 0.2 atmospheres. Figure 8.5 shows the results of these studies in a plot of the corrosion rates, k', against solution pH. The rates tend

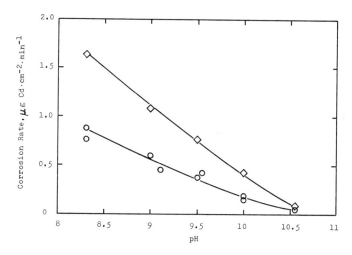

Figure 8.5. Rate of cadmium corrosion as a function of pH in carbonate and carbonate-free solutions. Experimental conditions: T = 25.5 ± 0.5°C, Ionic Strength = 0.01 ($NaClO_4$), 450 rpm, P_T = 1 atm, P_{O_2} = 0.2 atm, carbonate-free solution (upper curve), $[CO_3]_{TOT}$ = 10^{-3} M (lower curve).

to fall off with increasing pH, which is qualitatively consistent with a concurrent decrease in the equivalent cell potential (Equations 4 and 5) for the overall process, but may also have mechanistic implications. An asymptotic course for both curves in the higher pH region is also conspicuous. For the carbonate-free system the latter behavior may be related to an increased tendency for the removal of films by complexation with hydroxide. The formation of films in carbonate-free systems is indicated, particularly above pH 10 by a slight decrease in rates at an advanced stage of the reaction. This phenomenon will be given further consideration in the following section.

Effect of Chemical Corrosion Inhibitors

Carbonate species present in low concentrations substantially decelerate the rate of cadmium corrosion compared to carbonate-free systems. This effect is most pronounced in the lower pH range and tends to decrease in magnitude when pH increases (Figure 8.5). The gradual formation of a protective film on the cadmium surface is indicated by the fact that corrosion rates in carbonate systems exhibit a nonlinear behavior, that is, the rates are observed to fall off with increasing extent of the reaction. This effect is illustrated in Figure 8.6, showing the separate

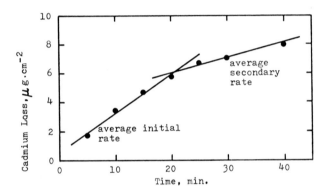

Figure 8.6. Corrosion of cadmium at pH 9.5 in the presence of 5×10^{-3} M total carbonate.

average slopes (rates) for the initial and secondary stages of the reaction. The tendency of rates to fall off is found to increase with increasing concentration of carbonate species in solution. This is shown in Figure 8.7 by a plot of the average initial and secondary corrosion rates at pH 9.5 as functions of carbonate concentration. The fact that corrosion inhibition by carbonate is quite strong even at pH 8.3 (Figure 8.5) and that a linear rate pattern for these experiments was observed suggests that film formation in these instances must be fast. Furthermore, and in agreement with cadmium carbonate precipitation data, this behavior suggests that film formation may either operate via

Aqueous Corrosion Chemistry of Cadmium 307

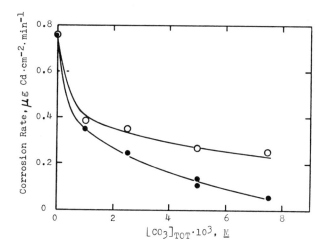

Figure 8.7. Effect of total carbonate concentration on cadmium corrosion rate. Experimental conditions: T = 25.5 ± 0.5°C, pH 9.5, Ionic Strength = 0.01 (NaClO$_4$), P$_{O_2}$ = 0.2 atm, 450 rpm. Upper and lower curve show average initial and average secondary rates, respectively.

combination with bicarbonate or carbonate ions since all of these reaction steps are considered rapid:

$$Cd^{2+} + OH^- \rightleftharpoons CdOH^+ \quad (17)$$

$$CdOH^+ + HCO_3^- \rightleftharpoons CdCO_{3(s)} + H_2O \quad (18)$$

$$Cd^{2+} + CO_3^{2-} \rightleftharpoons CdCO_{3(s)} \quad (19)$$

A similar reaction scheme can be used to describe the formation of Cd(OH)$_2$ films in carbonate-free systems. The tendency for corrosion rates to fall off during the reaction is much less pronounced and observable only at pH > 10 in these cases. Thus, the formation of Cd(OH)$_2$ films is expected to be similarly rapid.

Solidifying evidence in favor of carbonate film formation was obtained by treatment of the cadmium surface with solutions containing 10^{-1} M carbonate prior to corrosion experiments carried out in 10^{-3} M carbonate solutions. The corrosion rate following such treatment was found to be much reduced compared to the rate observed with an untreated surface (Figure 8.8).

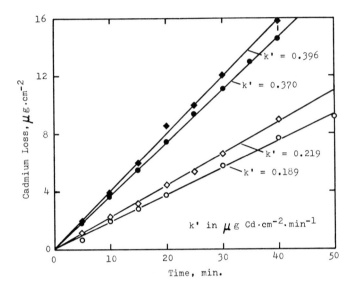

Figure 8.8. Effect of surface treatment with inorganic compounds on corrosion rate. Experimental conditions: T = 25.5 ± 0.5°C, pH 9.5, Ionic Strength = 0.01 (NaClO$_4$), P$_{O_2}$ = 0.2 atm, [CO$_3$]$_{TOT}$ = 10^{-3} M, 450 rpm.

	Conditions of surface treatment and further additives to testing medium	
Curve Designation	Pretreatment Conditions	Additive
solid circles	none	none
solid diamonds	15 min in solution of 1 g/l of (NaPO$_3$) x, pH 9.5	(NaPO$_3$)$_x$ 5 mg/l
open diamonds	15 min in solution of 1 g/l of Na$_2$SiO$_3$, pH 9.5	Na$_2$SiO$_3$ 5 mg/l
open circles	15 min in solution of 0.1 M NaHCO$_3$/Na$_2$CO$_3$, pH 9.5	none

In light of the present data, it appears that the different rate behavior found for carbonate and carbonate-free systems can be explained in terms of film permeability and/or film thickness, which in turn may be related to solubility phenomena.

Aqueous Corrosion Chemistry of Cadmium 309

In this connection it is noteworthy to mention that the extent of corrosion in carbonate systems was usually found to far exceed the solubility limits for cadmium, independent of a linear or nonlinear rate pattern. Since precipitation of cadmium on vessel walls or in solution was not observed, the rate decrease at an advanced stage of the reaction is therefore an indication of the progressing growth of the film, which is affected by the concentration of carbonate in solution.

Prevention of corrosion by addition of film-forming reagents is common practice, but is severely restricted to a few compounds in potable water treatment. The most natural approach of restraining corrosive attack of pipes is that of controlled deposition of calcium carbonate, the theory of which has been presented in a previous section. Other popular corrosion inhibitors used for protection of drinking water distribution systems are sodium metasilicate and sodium metapolyphosphates. The growth of protective layers of these compounds present in small concentrations is usually a slow process and cannot be easily simulated under laboratory conditions. Frequently, a high initial "shock" dosage followed by continuous application of small concentrations of these inhibitors is practiced. The feasibility of this method may be demonstrated on a laboratory scale.

Figure 8.8 shows the rates of cadmium corrosion for cases of surface pretreatment with carbonate, sodium metapolyphosphate and sodium metasilicate followed by exposure to small concentrations of these compounds during the corrosion experiment. The results indicate specific film formation for pretreatment with carbonate and metasilicate, whereas no such effect can be claimed for metapolyphosphates. Experiments carried out in the presence of small concentrations of metapolyphosphate as shown in Table 8.1 are consistent with this result. These experiments were performed at pH 9.5, ionic strength 0.01, 25.5 ± 0.5°C, and $[CO_3]_{TOT}$ of 10^{-3} M while mixing at 450 rpm.

While the corrosion-inhibiting properties of carbonate and silicate could be readily demonstrated, the apparent response to metapolyphosphates does not appear conclusive in view of the well-documented usefulness of these compounds for corrosion protection [5]. Kinetic effects and participation of calcium ions may play a major role in film formation by metapolyphosphates.

Table 8.1
Corrosion of Cadmium in the Presence of Sodium Metapolyphosphate

Calgonite ppm	Corrosion Rate k'
0	0.370
2.5	0.350
4.37	0.389
6.25	0.378

A more adequate examination of the role of metapolyphosphate and calcium carbonate in the corrosion of cadmium should include long-term exposure of actual galvanized pipe structures. In this context, it must be realized that corrosive attack of a minor component within an alloy structure, such as cadmium, is expected to decrease concurrently with a reduction of corrosion of the overall surface in the presence of protective coatings. Thus, compounds capable of coating the entire surface, without necessarily exerting metal-specific effects, should also be protective towards cadmium.

Corrosion with Galvanic Coupling

Polarization of cadmium by external coupling with a nobler metal, such as copper ($E° = +0.34$ volts) or brass, provides an additional driving force for the corrosive attack of cadmium. Thermodynamically, the potential difference of a Cu/Cd couple is markedly larger than that established by microscopic short-circuited galvanic couples (microcells or microcouples) on an inhomogeneous cadmium surface. As illustrated in Figure 8.9, the corrosion rate increases by a factor of about 10 when cadmium is galvanically coupled to copper. Consideration of the relative increase in surface area for oxygen reduction and the rates for the Cu/Cd and Cd systems suggest that the "free" surface area of cadmium under otherwise identical but "noncoupling" conditions is small compared with the geometric surface area employed.

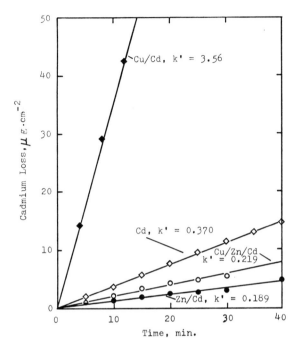

Figure 8.9. Effect of external galvanic coupling on rate of cadmium corrosion. Experimental conditions: T = 25.5 ± 0.5°C, pH 9.5, Ionic Strength = 0.01 (NaClO$_4$), [CO$_3$]$_{TOT}$ = 10^{-3} M; Surface areas: Cu = 6 cm^2, Cd = 3 cm^2, Zn = 4.5 cm^2.

This fact is additional evidence for the formation of surface films as proposed for the nongalvanic corrosion of cadmium in oxygenated aqueous systems. The galvanic experiment with copper was observed for 24 minutes. A gradual rate decrease was found (2.66 µg Cd cm^{-2}·min^{-1} average rate) beyond a time of 16 minutes, which again indicates formation of film during an advanced stage of the reaction.

Much, but not all, of the anodic role of cadmium is lost by coupling of the Cu/Cd system to the less noble zinc. This behavior is also indicated in Figure 8.9 (open circles) by the corrosion rate observed for a Cu/Zn/Cd system. Cadmium continues to corrode fairly rapidly when coupled to the less

noble zinc, that is, under conditions of cathodic protection.
This indicates that microcell-induced corrosion of cadmium may
proceed under favorable conditions, though to a lesser degree,
in spite of cathodic protection by zinc as an underlying process.

While measured on a quantitative scale, the galvanic experiments were performed to gain some qualitative insight into the corrosive behavior of cadmium under more complex conditions such as prevail in water distribution systems. The effect of relative areas of anodic and cathodic sites, oxygen accessibility, and other factors of influence on the corrosion rate must also be taken into account when considering more practical metal systems. Cadmium present within the microstructure of galvanized zinc may also become mechanically dislodged during the corrosion of zinc. In view of these possibilities, it is of interest to examine some properties of zinc relative to corrosion.

Zinc is less noble than cadmium ($E°$ = -0.76 volts) and dissolves readily in nonoxidizing acids. The solubility products for zinc hydroxide and zinc carbonate are 7 x 10^{-18} and 2.1 x 10^{-11}, respectively [9]. Thus, owing to the relative difference in these solubility products in comparison to cadmium, the solubility of zinc under most conditions of natural waters will be governed by pH rather than by the concentration of carbonate species. This suggests also that protective films on a zinc surface consist largely of $Zn(OH)_2$ or similar products but not likely zinc carbonate. Furthermore, zinc is less soluble than cadmium under a wide range of environments representative of drinking waters.

Although zinc is known to dissolve under alkaline conditions, this behavior is insignificant under the pH conditions of natural, and particularly drinking waters, in view of the magnitude of formation constants for zinc hydroxide complexes [10].

Consistent with the thermodynamic properties of zinc is its corrosive behavior in water as reflected by the data in Table 8.2 taken from Bhakalov, *et al.* [8]. Qualitatively, the corrosion properties of zinc resemble those of cadmium in the range pH 8 to 11, beyond which zinc exhibits less corrosion resistance. While the exact experimental conditions have not been cited, the data found for the corrosion of zinc indicate somewhat slower rates than obtained for cadmium in this study. At least, an agreement in the order of magnitude for corrosion rates of both metals is

Aqueous Corrosion Chemistry of Cadmium 313

Table 8.2
Corrosion of Zinc in Distilled Water as Function of pH

pH	Corrosion Rate $\mu g(Zn) \cdot cm^{-2} \cdot min^{-1}$
4	1.4
6	0.70
8	0.35
10	0.14
12	0.14
13	1.4

evident. Thus, the minimum rate for the corrosion of cadmium, present within a galvanized zinc surface, appears to be governed by the rate of corrosion of the less noble zinc, in which case the process may simply be one of mechanical dislodgement. Increased overall corrosion of both metals is very likely to occur when these are galvanically coupled to elements of lower corrosive potential, such as copper, brass, and passivated steel.

SUMMARY AND CONCLUSIONS

Cadmium may enter drinking waters by corrosion of galvanized pipes. Presumably this is a temporary problem in view of the present trend toward replacing galvanized steel by copper or even plastic plumbing materials. From the standpoint of the high toxic potential of cadmium to man [11], a study of the corrosion behavior of the metal in representative aqueous systems was felt of immediate importance in order to establish criteria for preventative measures.

The corrosion of cadmium in water between pH 8.3 and 10.55 was found to proceed under cathodic control via reduction of oxygen. The cathodic reaction is limited by oxygen transport to the corrosion site under a wide range of experimental conditions.

Corrosion rates were found to decrease substantially with increasing concentration of hydroxide and carbonate species. Precoating of the surface with carbonate or sodium metasilicate resulted in greatly reduced rates of attack relative to those when no pretreatment was provided. The data suggest that corrosion inhibition by carbonate, metasilicate, and hydroxide is attributable to the formation of protective films.

In spite of a film-forming tendency observed with solubility-governing ions, the corrosion of cadmium may lead to accumulation of the metal in solution far in excess of concentrations corresponding to calculated solubilities in these systems. With respect to drinking water contamination it seems irrelevant, however, whether the metal exists in a dissolved or dispersed solid (precipitated) state.

Inhibition of corrosion by pretreatment of cadmium metal with, or exposure to aqueous solutions containing small concentrations of sodium metapolyphosphate was not observed in this work. In view of the well-documented suitability of these compounds for corrosion prevention, the observed results are considered inconclusive. Long term experimentation, possibly on a pilot scale, seems necessary to establish more information on the usefulness of metapolyphosphates for this purpose.

Coupling of cadmium with the more noble copper was shown to result in drastic corrosion of the former, as can be expected thermodynamically. The high relative rate of corrosion of a Cu/Cd couple compared with rates observed for a pure Cd system is further evidence in favor of film formation for the case of nongalvanic attack of cadmium in oxygenated waters.

Under favorable conditions, "chemical" corrosion of cadmium coupled to less noble zinc may proceed at measurable rates in spite of cathodic protection by zinc. Otherwise, it appears that cadmium contained within galvanized zinc becomes mechanically dislodged during corrosion of zinc, which exhibits corrosive properties similar to those obtained for cadmium [8].

Corrosion of cadmium contained within galvanized zinc pipes may be minimized or prevented from occurring by:

1. Pretreatment of piping systems with solutions of carbonate or sodium metasilicate followed by low level feed of these compounds.
2. Manipulation and control of chemical factors (pH, alkalinity, hardness) so that the solubility of calcium carbonate

is exceeded in order to facilitate interior coating of the pipes with calcium carbonate.
3. Elimination of coupling with more noble metals (*e.g.*, copper, brass).

REFERENCES

1. Federal Water Pollution Control Administration. *Water Quality Criteria*. Report to the National Technical Advisory Committee to the Secretary of the Interior, Washington, D.C. (April 1, 1968).
2. Perry, H. M., I. H. Tipton, H. A. Schroeder, R. L. Steiner, and M. J. Cook. *J. Chron. Dis.*, 14, 259 (1961).
3. U.S. Public Health Service. *Public Health Service Drinking Water Standards*. U.S. Department of Health, Education and Welfare, Washington, D.C. (1969).
4. Chizhikov, D. M. *Cadmium*. (New York: Pergamon Press, 1966).
5. Butler, G. and H. C. K. Ison. *Corrosion and its Prevention in Waters*. (New York: Reinhold, 1966).
6. Bottles, D. G. *J. Amer. Water Works Assoc.*, 1, 55 (1970).
7. LaQue, F. L. and H. R. Copson. *Corrosion Resistance of Metals and Alloys*. (New York: Reinhold, 1963).
8. Bakhalov, G. T. and A. V. Turkovskaya. *Corrosion and Protection of Metals*. (New York: Pergamon Press, 1965).
9. Freiser, H. and Q. Fernando. *Ionic Equilibria in Analytical Chemistry*. (New York: Wiley, 1963).
10. Butler, J. N. *Ionic Equilibrium—A Mathematical Approach*. (Reading, Mass.: Addison Wesley, 1964).
11. Schroeder, H. A. *N. Eng. J. Med.*, 280, 836 (1969).
12. Weber, W. J. Jr. and H. S. Posselt. in *Aqueous-Environmental Chemistry of Metals*, A. J. Rubin, ed. (Ann Arbor, Mich.: Ann Arbor Science Publishers, 1974).

CHAPTER 9

SYSTEMATIC INVESTIGATION OF THE HYDROLYSIS AND PRECIPITATION OF ALUMINUM(III)

Phillip L. Hayden
Pollution Control Science Inc.,
Miamisburg, Ohio

Alan J. Rubin
Water Resources Center, College of Engineering
Ohio State University, Columbus, Ohio

Introduction	318
Purpose and Scope	318
Solubility of Metal Hydroxides	320
Aqueous Chemistry of Aluminum(III)	323
Experimental Methods and Materials	327
Preparation of Solutions	327
Precipitation Studies	328
Potentiometric Studies	330
X-Ray Diffraction Procedure	331
Studies on the Precipitation of Aluminum(III)	332
Determination of Precipitation Limits	332
X-Ray Diffraction Studies	341
Effect of Sulfate	343
Computer Analysis of Aluminum Hydrolysis	349
Computer Program SCOGS	349
Criteria Used for Analysis of Data	352
Evaluation of Hydrolysis Schemes	354
Final Evaluation of Data	359
Discussion and Conclusions	363
Determination of Equilibrium Constants	363
Structure and Form of Aluminum Hydroxide	372
Effect of Anions on Precipitation	375
Inferences to Chemical Coagulation	377
Summary of Hydrolysis Constants	378

INTRODUCTION

Purpose and Scope

Aluminum salts have been used extensively for the clarification of raw water for drinking purposes, in the treatment of industrial wastewater and recently in the chemical treatment of municipal sewage. These aluminum salts are commonly called coagulants and the process is generally known as "coagulation." Although the coagulation of various colloidally dispersed suspensions has been investigated and reported in the literature, with few exceptions most of these studies have been somewhat empirical and the results ambiguous. Because many researchers disagree over the exact nature of the hydrolyzed aluminum, it is not surprising that there is also not much agreement over the role of aluminum in coagulation processes, especially as to whether the active hydrolysis products are soluble or insoluble. This disagreement is quite understandable in that the aqueous reactions are very complicated and difficult to define.

In addition, there are semantic inconsistencies in the terms used to describe aluminum systems. The word "precipitation" is one such term that is often used ambiguously. Several researchers, when discussing the aluminum system, have stated that precipitation occurs when a sediment collects in the bottom of the vessel. However, most chemists, especially analytical chemists [1,2] view precipitation as the formation of a solid phase when the product of the active masses of the reactants exceeds the solubility product. They recognize that although precipitation of an insoluble phase occurs, a settleable recoverable precipitate may not always form. Sometimes the precipitate is colloidally dispersed either during its formation because of adsorption of potential determining ions or through peptization during washing or dilution. Colloid chemists sometimes refer to the settling of sols after coagulation as precipitation. However, for the purposes of this paper, precipitation will be defined as the formation of an insoluble phase from soluble reactants as a result of their active masses or concentrations exceeding that predicted by the solubility product, but does not refer to the settleability of that phase.

Hydrolysis and Precipitation of Aluminum(III) 319

In some cases, the formation of a solid phase does not always occur when the active masses of the reactants or the ion product exceed that predicted by the solubility product. In this case, the solution is said to be oversaturated or metastable with respect to formation of a solid phase. A metastable solution may persist over a long period of time even though precipitation has occurred because the solid phase has not evolved into its most ordered or inactive state. A solution cannot be considered stable until the ion product equals the solubility product for the most inactive phase for that particular pH, temperature, and ionic strength. If any of these conditions are changed, the solution will be metastable until true heterogeneous equilibrium is reestablished. Metastable solutions are often studied in actual practice because metastable conditions generally prevail.

The stability of a solution, which refers to a state or condition of that solution with respect to formation of the most insoluble or inactive solid phase, must be carefully distinguished from the colloidal stability of the precipitate. A precipitate which is dispersed as a sol is colloidally stable and may not settle even after long periods of time. A precipitate, which is not colloidally dispersed and therefore would settle if allowed to stand undisturbed, is generally referred to as unstable.

Because the heterogeneous reactions of aluminum have not been well defined, especially the solubility and colloidal stability of the condensed phase in metastable solutions, it is difficult to develop a rational model of "coagulation" using aluminum salts. Therefore, the purposes of this research were to examine the aqueous reactions of aluminum(III) over a wide range of pH and metal ion concentrations and to develop a rational model of the aqueous chemistry of aluminum(III). Specific objectives were to identify hydrolyzed species and determine their equilibrium constants. Both soluble and insoluble systems were examined and solubility products for aluminum hydroxide were also determined. Studies with freshly prepared and aged solutions involved light scattering techniques, X-ray diffraction, and computer analysis of potentiometric data.

Solubility of Metal Hydroxides

The solubility of a metal hydroxide such as that of aluminum is generally written as a dissociation reaction in the following manner:

$$Me(OH)_n(s) \rightleftharpoons Me^{n+} + nOH^- \qquad (1)$$

so that

$$K_{s_0} = \{Me^{n+}\}\{OH^-\}^n \qquad (2)$$

where K_{s_0} is the thermodynamic solubility product and the braces indicate activities. The reaction can also be written as hydrolysis:

$$Me^{n+} + nH_2O \rightleftharpoons Me(OH)_n(s) + nH^+ \qquad (3)$$

and as before:

$$*K_{s_0} = \frac{\{H^+\}^n}{\{Me^{n+}\}} \qquad (4)$$

If the metal hydroxide were formed by the simple combination of one metal ion and n hydroxides, both of these expressions could adequately explain the solubility as a function of pH. But since monomeric and polynuclear hydrolyzed species form both in the acid and alkaline pH ranges, these must be considered in order to calculate the true solubility. The solubility of the metal hydroxide is therefore the sum of the concentrations of all the soluble species:

$$\{Me\}_{soluble} = \{Me^{n+}\} + \sum_{m}\sum_{q} m\{Me_m(OH)_q^{(nm-q)+}\} \qquad (5)$$

and substituting

$$\{Me\}_{soluble} = \{Me^{n+}\} + \sum_{m}\sum_{q} \frac{p\beta_{m,q}\{Me^{n+}\}^m}{\{H^+\}^q} \qquad (6)$$

where $\beta_{m,q}$ is the overall formation constant for the generalized species $Me_m(OH)_q^{(nm-q)+}$

In the presence of the precipitate the concentration of the free metal ion Me^{n+} is determined by the solubility product.

Hydrolysis and Precipitation of Aluminum(III) 321

Therefore Equation (4) can be solved for Me^{n+} and substituted into Equation (6)

$$\{Me\}_{soluble} = \frac{\{H^+\}^n}{*K_{s_0}} + \sum_m \sum_q \frac{{}^*\beta_{m,q} \{H^+\}^{nm-q}}{*K_{s_0}^m} \tag{7}$$

If all the hydrolyzed species and the formation constants are known, the true solubility of the metal hydroxide can then be calculated using Equation (7).

If the ionic species in equilibrium with the precipitate are not known, they can be postulated by analysis of experimental solubility data. Basically, this approach consists of determining the solubility of the metal hydroxide at various concentrations as a function of pH and constructing a log-log solubility diagram by plotting log concentration soluble metal against pH. If the solubility curve or precipitation boundary is curved, indicating that there is not a single predominant species in equilibrium with the precipitate over the concentration range under consideration, Equation (7) can be applied. In general, various complex species are postulated and, with the aid of a computer, the constants of the postulated species are refined by least squares adjustment until the difference between the calculated solubility and experimental solubility is at a minimum. In this way the structure and equilibrium constants of the hydrolyzed species can be established. This approach has been reported by Bilinski, Branica and Sillén [3] for zirconium hydroxide using a computer program "LETAGROP" developed by Sillén and Ingri [4].

If the solubility curve is linear over the concentration range under consideration, then an assumption can be made that there is but one predominant species in equilibrium with the precipitate. This approach, which has been used extensively by Težak and co-workers, has been summarized by Furedi [5]:

$$mMe(OH)_n(s) + (nm-q)H^+ \rightleftharpoons Me_m(OH)_q^{(nm-q)+} + (nm-q)H_2O$$

$$K_{s_{mq}} = \frac{\{Me_m(OH)_q^{(nm-q)+}\}}{\{H^+\}^{nm-q}} \tag{8}$$

Taking logarithms and rearranging into a linear form of
$y = ax + b$:

$$\log\{Me_m(OH)_q^{(nm-q)+}\} = -(nm-q)pH - pK_{s_{mq}} \qquad (9)$$

If there is but one species in equilibrium with the precipitate then it can be assumed that the total concentration of the metal is the concentration of the complex. Therefore, the concentration of the hydrolyzed species is equal to the total soluble metal concentration C_{Me} divided by m, the number of metal ions per complex

$$\log C_{Me} = -(nm-q)pH - (pK_{s_{mq}} - \log m) \qquad (10)$$

Therefore, by plotting $\log C_{Me}$ against pH, the charge of the hydrolyzed complex can be determined from the slope (nm-q) and the solubility constant from the intercept.

However, there are several limitations to this technique of analyzing precipitation boundaries or pH limits of solubility. First of all, the experimental boundary must represent or closely approximate the true solubility curve. Under most experimental conditions, the precipitation boundary represents the limits of a metastable solution assuming homogeneous precipitation, whereas, the true solubility curve represents a saturated solution. If there is a substantial difference between a metastable and a saturated solution, the experimental solubility constant will indicate that the precipitate is more soluble than it actually is. In a similar fashion, especially with very insoluble salts, localized precipitation which may occur during mixing will shift the precipitation boundary in the opposite direction indicating the solid phase is more insoluble. But the effects of both localized precipitation and supersaturation can be minimized if the system is at equilibrium when the solubility limits are measured. The primary limitation to this approach for determining hydrolyzed metal complexes is that it is impossible to distinguish between monomers and polynuclears with identical charge.

Aqueous Chemistry of Aluminum(III)

The extremely complex and diverse nature of the aqueous chemistry of aluminum is reflected both in the past and present literature. The hydrolytic reactions of aluminum have been studied by many investigators of various disciplines for many purposes. Although there are many areas of agreement, there are also several of disagreement. The principal area appears to be whether the hydrolyzed aluminum species are monomeric or polynuclear. Part of this ambiguity must be attributed to the fact that aluminum solutions are very slow to approach a true equilibrium state so that meaningful measurements are difficult to obtain.

When aluminum salts are dissolved in water in the absence of complexing anions, the free metal ion Al^{3+} first hydrates, coordinating six water molecules in an octahedral orientation [6], and then reacts forming various hydrolytic species. The first step in hydrolysis has been assumed by many to proceed in the following manner:

$$Al(H_2O)_6^{3+} + H_2O \rightleftharpoons Al(H_2O)_5OH^{2+} + H^+ \qquad (11)$$

and the thermodynamic equilibrium constant for this reaction is given by

$$*K_1 = \frac{\{AlOH^{2+}\}\{H^+\}}{\{Al^{3+}\}} \qquad (12)$$

where $\{AlOH^{2+}\}$ and $\{Al^{3+}\}$ are the activities of $Al(H_2O)_5OH^{2+}$ and $Al(H_2O)_6^{3+}$, respectively, the waters of hydration being omitted for simplicity. This reaction has been studied primarily by dissolving pure aluminum salts in water and diluting to various concentrations. Using either potentiometric and/or conductimetric techniques, the concentrations of the reacting species were determined and the constant calculated. There appears to be good agreement on the value of $*K_1$ of approximately 10^{-5} at zero ionic strength as listed in Table 9.1.

The second step in monomeric hydrolysis, which involves formation of the dihydroxo-aluminum(III) species $Al(OH)_2^+$, is

324 Aqueous-Environmental Chemistry of Metals

Table 9.1

Summary of Hydrolysis Constants for $AlOH^{2+}$

Source	$p*K_1$	Temperature (°C)
Bronsted and Volquartz [7]	4.89	15
Hartford [8]	4.96	25
Schofield and Taylor [9]	4.98	25
Ito and Yui [10]	5.10	25
Kubota [11]	5.03	25
Frink and Peech [12]	5.02	25
Raupach [13]	5.00	25
Hem and Roberson [14]	4.75	25

not so well documented. Most of the evidence for monomeric hydrolysis, which was first postulated by Bronsted and Volquartz [7], has been gained from solubility studies. Typical of this approach, Raupach [13] equilibrated various aluminum solid phases [freshly prepared $Al(OH)_3(s)$, gibbsite, diaspore, and bauxite] in solutions of varying pH. The solutions were adjusted to 0.01 M with potassium sulfate because colloidal materials were observed to remain in solution after centrifugation using the same concentration of potassium chloride. The residual or soluble aluminum was determined after separation from the solid phase and was plotted on a graph as a function of pH. The solubility curves were compared with theoretical curves which were calculated assuming the monomers $AlOH^{2+}$, $Al(OH)_2^+$ and $Al(OH)_4^-$. A similar solubility study has been reported by Gayer, Thompson and Zajick [15] for aluminum hydroxide. They determined the solubility constants for Al^{3+}, $AlOH^{2+}$, $Al(OH)_2^+$ and $Al(OH)_4^-$ in equilibrium with a crystalline phase, which according to Hem and Roberson [14] was bayerite. Using a somewhat different approach, Reesman [16] has also interpreted his solubility results for aluminum-bearing minerals only in terms of monomeric hydrolysis. He measured the migration of aluminum ions at different pH on chromatographic paper strips at an applied voltage of 800 volts and concluded that the principal aluminum ions over the pH range 3.0 to 9.0 were $Al(OH)_4^-$ and $Al(OH)_2^+$.

There are numerous investigators who have not been able to interpret their data by assuming simple monomeric aluminum hydrolysis. Since 1908, when Bjerrum first detected polynuclear chromium(III) complexes, many other metals have been shown to form polymeric species. Over the years the principal group of investigators who have studied metal ion hydrolysis have been Sillén and his Swedish co-workers [17,18]. They have developed many mathematical and graphical techniques to analyze the complex metal hydrolysis equilibria and calculate equilibrium constants [4,19,20]. Most of their studies have been conducted with aqueous solutions at 25°C and at constant ionic strength (usually 3 M $NaClO_4$). They have interpreted their data primarily by assuming either a single polynuclear complex or a series of polynuclears with a basic "core" and various numbers of attached "links." One of the more extensive potentiometric investigations into the aqueous chemistry of aluminum(III) was conducted by Brosset [21], a member of this group. He originally analyzed the data by assuming an infinite series of complexes of the form $Al[(OH)_3 Al]_n^{3+}$ in the acid range and $[Al(OH)_3]_n OH^-$ in the alkaline range. However, two years later, Brosset, Biedermann and Sillén [22] reviewed the same data and concluded with reservations that a single complex with hexameric structure, $Al_6(OH)_{15}^{3+}$, was the principal species in acid solutions and that the aluminate ion is predominant in alkaline solutions. They did not rule out other conceivable complexes such as $Al_6(OH)_{20}^{4+}$ but since the hexamer was similar in structure to crystalline gibbsite, $Al_6(OH)_{15}^{3+}$ was a logical product of aluminum hydrolysis. In a later paper Biedermann [18] described a study of aluminum(III) hydrolysis conducted at 50°C to accelerate hydrolysis reactions. His results indicated a septamer $Al_7(OH)_{17}^{4+}$ and a large polycation, $Al_{13}(OH)_{34}^{5+}$. Biedermann also concluded, however, that even though the studies were conducted at 50°C, the uncertainty of the data was high because of the nonequilibrium conditions.

Kubota [11] and Faucherre [23] found evidence of a dimer $Al_2(OH)_2^{4+}$ in acid solutions. Faucherre suggested that at aluminum concentrations greater than 0.01 M the dimer formed, and a monomer formed below 0.005 M. Aveston [24] also postulated a dimer in aluminum perchlorate solutions using both equilibrium ultracentrifugation and potentiometric techniques. Analysis of the potentiometric data with a computer program employing a least squares approach indicated both a dimer and a large polycation, $Al_{13}(OH)_{32}^{7+}$. Using equilibrium constants refined by the computer program, the degree of polymerization was calculated and agreed with the ultracentrifugation results.

Matijević and co-workers have also studied aluminum(III) hydrolysis [25-28] and complexation [29,30]. The concentration of various ionic metal species required to coagulate and restabilize sols was correlated to the charge and, hence, to its extent of hydrolysis or complexation. By analyzing the boundaries between concentration regions of coagulation and restabilization, the ratio of ligand to metal ion was deduced. Using criteria such as this, Matijević, Mathai, Ottewill and Kerker [27] suggested the octamer $Al_8(OH)_{20}^{4+}$ as the principal hydrolyzed species in aqueous aluminum solutions.

The initial products of aluminum precipitation are amorphous with no definite ordered structure, but with aging they evolve into identifiable crystalline phases. At normal temperature, gibbsite forms very slowly in the acid pH range. Mixtures of pseudoboehmite, bayerite and possibly nordstrandite form in the alkaline range. A review by Schoen and Roberson [31] discusses the formation of these crystalline phases of aluminum hydroxide in detail. Papers by Hsu [32-34], and Hem and Roberson [14] provide a good background. The values of solubility constants of the various phases are listed in Table 9.2.

The effects of various anions on the precipitation of aluminum(III) and other hydrolyzing metals have been studied extensively by Thomas and co-workers (e.g., References 35-37). Their work has been reviewed by Pokras [38-40]. They found that various anions significantly affected the solution pH, form and nature of the aluminum precipitate, and pH of maximum precipitation. The order of stability of aluminum(III) complexes was reported to be oxalate > sulfate > chloride > bromide > iodide > nitrate. The rate of solution of hydrous alumina in various acids was found to follow the same order except that phosphoric acid was more effective than oxalic acid. They found

Table 9.2

Summary of Solubility Constants for Aluminum Hydroxide

Phase	log K_{S_O}	log $*K_{S_4}$	Reference
Pseudoboehmite	-32.90 (20°C)	-12.74	[41]
	-31.72 (30°C)	-12.87	[41]
		-12.5	[22]
		-12.70	[14]
Bayerite	-32.95 (25°C)	-14.53	[15]
		-13.95	[14]
		-13.84	[42]
Gibbsite	-32.65 (25°C)		[14]
	-33.51 (25°C)		[43]

that the stability of the aluminum(III) complex significantly affected the pH of maximum precipitation of aluminum hydroxide depending on whether the anion was incorporated into the precipitate or displaced by hydroxide ion. Heat and solution age also had a profound effect on the properties of aluminum hydroxide sols. Marion and Thomas [35] proposed that aluminum polymerized through a bridging mechanism which they termed "olation" where two aluminum ions are bound (bridged) by two hydroxide ions. As the polymers age, hydrogens split off from the bound hydroxides resulting in bridging by oxygens, or "oxolation." The oxolation-olation bridging mechanism accounted for the observed decrease in the pH of the aluminum solutions.

EXPERIMENTAL METHODS AND MATERIALS

Preparation of Solutions

Solutions were prepared by dispersing reagent grade chemicals in double-distilled deionized water and stored in polyethylene bottles. All solutions were filtered twice through a 0.45-micron Millipore membrane filter to reduce their turbidity to a low constant value.

Stock acid solutions were prepared by diluting concentrated nitric acid and were standardized periodically against tris(hydroxymethyl)aminomethane, which had been dried at 105°C and stored in a vacuum desiccator, using methyl purple as the indicating dye. Stock base solutions were prepared by dispersing reagent sodium hydroxide pellets in distilled water, heating overnight and filtering to remove any insoluble carbonates. These solutions were standardized against nitric acid. Aluminum(III) stock solutions were prepared in concentrations greater than 0.10 M to prevent hydrolysis and subsequent aging of the solutions [25]. The concentration of each aluminum solution was determined by alkalimetric titration or colorimetrically using eriochrome cyanine R as the chromogenic reagent. The latter procedure was modified slightly in that the samples were buffered at pH 5.5 instead of 6.0 [44]. This adjustment of pH resulted in a more sensitive measurement of the aluminum concentration. The molar absorptivity of the aluminum-ECR complex was calculated to be 67,500 at this pH.

All pH measurements were made using Sargent-Welch model DR or model NX pH meters with Sargent combination electrodes. The pH meters were calibrated with buffers before, during and after each experiment. If the instrument required adjustment during a set of measurements, the pH of each sample was redetermined.

Precipitation Studies

The turbidities of the aluminum solutions were estimated using 90° light scattering measurements. A Coleman model 9 nephelometer with unfiltered white light was used for the more concentrated aluminum solutions in combination with a Coleman solid state regulated D.C. power supply with three voltage settings. Although the nephelometer was designed to be operated at 6.2 volts, operation at 7.2 and 8.5 volts expanded the useful range of measurements. The turbidity measurements are expressed as "relative scattering" units on a 0 to 100 scale. Prior to each experiment, round 19 x 105-mm Coleman cuvettes were matched to within 0.5 relative scattering units using distilled water. The

Hydrolysis and Precipitation of Aluminum(III) 329

nephelometer was calibrated before each set of measurements using Coleman turbidity standards at a power setting of 7.2 volts. For the very dilute aluminum solutions, light scattering determinations were made using a Brice-Phoenix Light Scattering Photometer, Universal 1000 series, in accordance with the directions detailed in the operation manual OM-1000A. A large cylindrical cell was used for all the measurements (Brice-Phoenix model C-101). The cell was cemented to a small rectangular glass plate which fit exactly in a slot in the rotating arm of the instrument and aligned so the flat entrance window of the cell was precisely perpendicular to the incident light beam. The alignment was checked by observing the image of light reflected back on the collimating tube diaphragm. The cell was properly aligned as maximum transmittance was observed at an angle of 0°. The reported values are the average of three separate measurements using blue light at a wavelength of 436 nm. Both the Coleman nephelometer and Brice-Phoenix scattering photometer gave identical results with solutions at intermediate aluminum concentrations.

For each experiment, the concentration of aluminum(III) was held constant and the pH systematically varied by adding sodium hydroxide. Samples for the more turbid aluminum solutions which could be analyzed with the Coleman nephelometer were prepared by pouring the contents of glass vials containing the exact amounts of base and distilled water into round cuvettes containing aliquot portions of aluminum solution. The concentration of the aluminum reported is the total concentration in the final 15-ml sample. The cuvettes were stoppered, mixed for 20 seconds and placed in the nephelometer for the first reading. The cuvettes were then removed, the pH was measured, followed by storage without further agitation until time for the next reading.

Samples for the aging studies and the less turbid aluminum solutions were prepared by diluting aliquot portions of aluminum and exact amounts of base to 100 ml in a volumetric flask. The samples were mixed thoroughly and transferred to 4-oz plastic bottles. The plastic bottles were placed in an Eberbach shaker and agitated vigorously at 200 rpm for 15 minutes and then transferred to the constant temperature shaker bath thermostated at 25°C and shaken at 80 rpm. The samples were removed periodically for pH and turbidity measurements. For the turbidity measurements, the samples were allowed to stand undisturbed for one hour before 15 ml were withdrawn from the top with a pipette and transferred to the cuvettes. For the samples aged only one hour, only about ten minutes were allowed for settling.

Potentiometric Studies

Because aluminum solutions are slow to approach equilibrium upon adding base, a serial pH titration technique was employed. A series of samples was prepared in 4-oz plastic bottles which had been rinsed in 1:1 nitric acid, scrubbed with Alconox detergent, rinsed several times with distilled water, and allowed to air dry. Aliquot portions of aluminum nitrate, sodium nitrate and excess nitric acid were pipetted into 100-ml volumetric flasks, diluted to the mark and transferred to the plastic bottles. The aluminum concentrations studied were between 1.0×10^{-4} M and 1.0×10^{-3} M, above pH 3 and below the pH where precipitation was first detected by light scattering. Nitric acid was added to the samples to lower the pH so that only the free metal ion was present initially, and if any acid was present in the aluminum nitrate salt, the amount added was much larger than that originally present. Addition of the acid also minimized the adsorption of aluminum onto the glass surface of the volumetric flask during preparation. Sodium nitrate was added to adjust the final ionic strength to 0.15. Exact amounts of sodium hydroxide were added with a 2-ml microburet while the samples were swirled to insure mixing. The samples were placed in an Eberbach shaker-bath thermostated at 25°C and shaken at 80 rpm. The samples were removed periodically and the pH measured. While the pH of the samples was being measured, nitrogen gas was bubbled into the solution after passing through a series of scrubber bottles containing pyrogallol to remove oxygen, saturated lime solution to remove carbon dioxide, and distilled water. Bubbling of the nitrogen through the samples insured mixing of the solutions and exclusion of carbon dioxide. Light scattering measurements were

Hydrolysis and Precipitation of Aluminum(III) 331

made to check for the presence of precipitate and only the clear samples were retained for analysis. After the solutions had aged three months at 25°C, some of the sample bottles were heated five days at 65°C, cooled to 25°C, and the pH was measured. Approximately 25 samples were prepared at each of three concentrations along with a series of duplicates at 1.0×10^{-3} M.

X-Ray Diffraction Procedure

Samples for X-ray diffraction analysis were prepared by titrating 500 ml of 0.0378 M aluminum nitrate and 0.30 M sodium nitrate (if used) with 0.910 M sodium hydroxide. The base was added very slowly over a period of 30 minutes to the aluminum solutions which were vigorously mixed on a magnetic stirrer. Nitrogen gas was bubbled through the solution to exclude carbon dioxide. The pH was adjusted so that a colloidal suspension of aluminum hydroxide would form in two sets of samples and a settleable precipitate would form in two other sets of samples. A fifth set of samples was prepared with 0.30 M sodium nitrate to determine the effects of a high concentration of salt on the X-ray diffraction patterns.

After the final adjustment of pH, the solutions were transferred to 16-oz plastic bottles. The solutions were "aged" in two ways. Most of the sample bottles were placed in an Eberbach constant temperature bath maintained at 25°C and shaken at 80 rpm. The rest of the samples were placed in a constant temperature bath at 65°C. The sample bottles were removed at specific intervals, the pH measured and precipitates recovered by centrifuging 50-ml aliquots of the solution in a Sorval Model SS4 Manual Superspeed Centrifuge at 10,000 rpm for 10 minutes. In those solutions where the precipitate was colloidally dispersed, the precipitate was recovered by adding enough sodium nitrate to coagulate the sol followed by centrifugation. The precipitate was transferred from the centrifuge tubes and dried in a vacuum desiccator over calcium sulfate (Drierite). The precipitates were extensively hydrated and decreased over 95% in volume on drying. The dried precipitates recovered from the acid solutions at an hydroxide per aluminum ratio of less than 3.0 were hard, brittle and transparent. Precipitates recovered from solution at greater hydroxide per aluminum ratios were softer and milky white in color. All the dried precipitates were ground with a mortar and pestle to a powder having the consistency of flour. The ground samples were mounted in an aluminum holder and analyzed with a Norelco high-angle X-ray diffractometer using Cu-Kα (Ni filtered) radiation.

STUDIES ON THE PRECIPITATION OF ALUMINUM(III)

Determination of Precipitation Limits

The solubility and colloidal stability of aluminum(III) hydroxide precipitates were studied using light scattering techniques. Basically, with each experimental series, the critical pH limits of precipitation and dissolution were determined by obtaining curves of scattering intensity as a function of pH at a given aluminum salt concentration. Typical results at 1.0×10^{-3} M aluminum nitrate obtained with the Coleman nephelometer are shown in Figure 9.1. The formation of condensed phase in acid solutions is indicated by a sharp increase in turbidity which occurs over a very narrow pH range. By extrapolating the data points back to the initial scattering values, the pH limit of precipitation can be determined. This pH limit, designated as the pH_p, is the critical hydrogen ion concentration at which condensed phase just begins to form. Below this pH no solid phase is detected. In alkaline solutions, as the pH is increased, a sharp decrease in scattering again occurs over a narrow pH range and the pH of dissolution can be determined in a similar fashion. This limit, designated as the pH_d, is the critical hydrogen ion concentration at which the precipitate is just completely dissolved. The pH range over which a given concentration of aluminum(III) is insoluble at a specified time is defined by the pH_p and the pH_d. Thus, for 1.0×10^{-3} M aluminum nitrate the pH range of precipitation at one hour was experimentally determined to be 4.55 to 9.25.

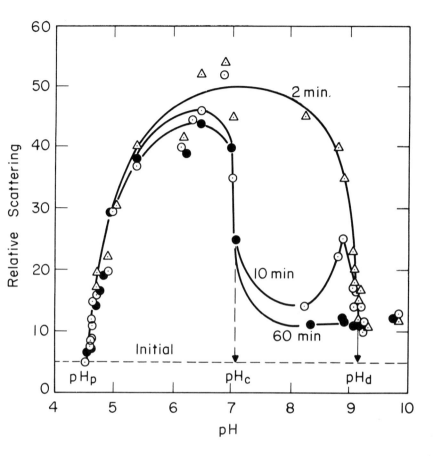

Figure 9.1. Typical turbidity-pH curves obtained with Coleman nephelometer for fresh solutions of 1.0×10^{-3} M $Al(NO_3)_3$.

The precipitate was colloidally dispersed in acid solutions. Above neutrality, as indicated by the 10- and 60-minute scattering curves in Figure 9.1, the precipitate settled rapidly. The boundary between the stable and settleable precipitate is designated as the pH_c or pH of coagulation and was determined by extrapolation back to the initial scattering value. The solutions in the acid pH range where the stabilized precipitate formed were perfectly clear to the naked eye, but a well defined Tyndall beam was observed when placed in the nephelometer.

Figure 9.2 shows typical results obtained with the Brice-Phoenix Light Scattering Photometer. Generally, this instrument was used for aluminum concentrations below 5.0×10^{-4} M which was about the lower limit of sensitivity for the Coleman nephelometer. The results are in good agreement with previous results

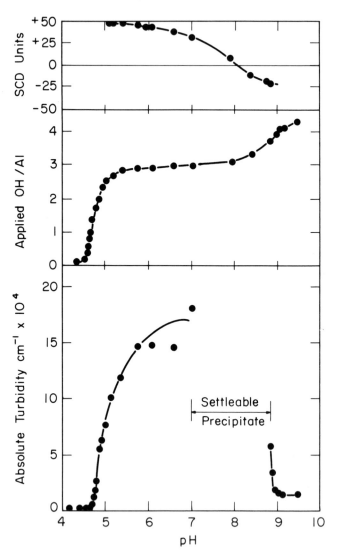

Figure 9.2. Comparison of OH/Al ratios and streaming current data with light scattering curve of fresh solutions of 5.0×10^{-4} M Al(NO$_3$)$_3$.

at higher aluminum concentrations. A stabilized precipitate formed in the acid range up to about pH 7 and a flocculent precipitate formed above neutrality. Light scattering measurements of the large flocculent particles were impossible because of the sensitivity of the instrument to the particle size using blue light at 436 nm. The flocculent aluminum hydroxide, which had originally settled in the plastic bottles used for preparation of the samples, were redispersed when the solutions were transferred to the scattering cell. The boundaries were well defined, however, and the limits of precipitation were pH 4.69 to 8.95 at 5.0 x 10^{-4} M aluminum nitrate.

Also shown in Figure 9.2 is a plot of applied OH/Al ratio as a function of pH. This relationship indicates that in freshly prepared solutions, a condensed phase was initially detected after one hydroxide per aluminum had been added and the precipitate dissolved after four hydroxides had been added. The pH_c of 7.1 corresponded to an OH/Al ratio of approximately 3, which can be considered the equivalence point in the titration of aluminum by sodium hydroxide. In addition to light scattering data and applied OH/Al ratios, Figure 9.2 shows measurements obtained with a Waters Associates Streaming Current Detector as a function of pH. Streaming current data indicates that aluminum hydroxide hydrosols are strongly positively charged and that the precipitate in the alkaline pH range was weakly negative. A zero point of charge (ZPC) at approximately pH 8 was obtained by connecting a smooth curve through the experimental points and determining the pH corresponding to zero SCD units. Although it is uncertain what is the exact electrochemical property of the precipitate that is being measured by streaming current, the ZPC obtained does appear logical in that pH 8 corresponds to the middle of the pH range where the settleable precipitate forms. This pH is consistent with results determined by other methods [15,45].

Figure 9.3 shows the light scattering characteristics of 1.0 x 10^{-3} M aluminum nitrate solutions aged 24 hours and 1 week compared to fresh solutions measured 1 hour after preparation. The 1-hour scattering curve is typical of freshly prepared aluminum solutions and the limits of precipitation are almost identical to Figure 9.1. However, light scattering measurements

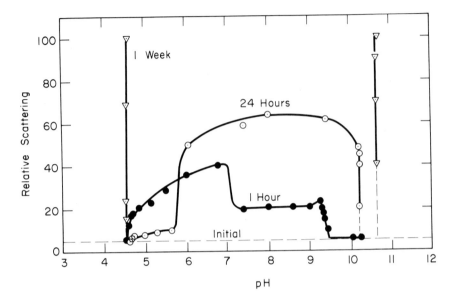

Figure 9.3. Effect of solution age on the pH of precipitation. 1.0×10^{-3} M Al(NO$_3$)$_3$ (25°C, 80 rpm).

of aluminum solutions aged 24 hours are entirely different. The turbidity of solutions in the acid pH range up to pH 6 decreased and the pH of precipitation is not well defined. The decrease in light scattering intensity is apparently due to dissolution of particles formed during preparation. The increase in turbidity above pH 6 is due to a breaking up of the flocculent settleable precipitate into small particles which are dispersed into solution because of the shaking action. The pH$_d$ or pH of precipitate dissolution shifted to higher pH with aging time, thus widening the pH range of precipitation. The pH boundary between the stable sol and settleable precipitate or the pH$_c$ at 24 hours was determined to be pH 5.90, compared to a pH of 7.0 at one hour. The change of both the pH$_c$ and pH$_d$ of the aluminum hydroxide precipitate after 24 hours of aging was due to a change in solution pH. For example, the pH of a solution prepared with an applied hydroxide per aluminum ratio of 3.0 was determined to be 6.80 at 1 hour and 5.90 after 24 hours of aging. Similar shifts in pH were

Hydrolysis and Precipitation of Aluminum(III) 337

observed in those solutions prepared in the alkaline range, except the pH increased with age.

In Figure 9.3 only part of the light scattering-pH curve defining the limits of precipitation for aluminum nitrate solutions aged one week is shown because the solutions in the intermediate pH ranges were too turbid to be measured. The pH range of precipitation measured at one week was well defined and the scattering boundaries were sharp with respect to pH. For this particular set of data, it is coincidental that the pH_p at one week was almost identical to that determined at one hour.

The turbidity data do not clearly indicate the boundary between the colloidal aluminum sol and settleable precipitate, but this limit was an OH/Al ratio of 3.0. The difficulty in obtaining turbidity data to support this value resulted from the fact that the flocculent precipitate, which formed in the range OH/Al of 3.0 to 4.0, was dispersed by shaking during aging for periods of 24 hours and longer; even after the samples were allowed to stand undisturbed for one hour, the solutions were still turbid. However, when the samples were viewed in the nephelometer, there was a distinct difference between the opaque solutions of the colloidal sols and the corase suspensions of small discrete particles.

Although the OH/Al ratios corresponding to the limits of coagulation and precipitate dissolution did not change with aging, the OH/Al ratio at which precipitation was initially detected did change. At 1 hour, this ratio was 1.0, at 24 hours it was 2.0, and at 1 week it was 0.75. Apparently the shift from 1.0 to 2.0, 24 hours after initial preparation, is due to dissolution of particles or localized precipitate formed during mixing. The shift from 2.0 to 0.75 after one week indicates a solution proceeding from a highly oversaturated state to a saturated or equilibrium condition.

The pH limits of solubility and stability of the aluminum hydroxide precipitate determined from the results shown in the figures and similar data are summarized in Table 9.3. These data were also plotted against pH as a function of the logarithm of the applied aluminum(III) concentration, as shown by Figure 9.4. One-hour data are represented by circles, 24-hour data by

Table 9.3
Summary of Critical pH Limits of Precipitation in Aluminum Nitrate Solutions

Molar Conc. $Al(NO)_3$	log M	pH_p (1 hr)	pH_p (24 hr)	pH_p (3 mo)	pH_p (Heated*)	pH_c (1 hr)	pH_d (1 hr)	pH_d (24 hr)
5.0×10^{-2}	-1.30					5.80		
4.0×10^{-2}	-1.40	4.05					10.90	
3.3×10^{-2}	-1.48	4.10					10.80	
3.0×10^{-2}	-1.52		4.22					
2.7×10^{-2}	-1.58					6.05		
2.0×10^{-2}	-1.70	4.17					10.60	11.47
1.7×10^{-2}	-1.77					6.30		
1.5×10^{-2}	-1.82	4.23	4.32				10.35	
1.0×10^{-2}	-2.00	4.30	4.40	3.98	3.92		10.25	11.15
5.0×10^{-3}	-2.30	4.40	4.51	4.05		6.83	9.90	10.84
3.0×10^{-3}	-2.52				4.16			
1.0×10^{-3}	-3.00	4.55	4.68	4.25	4.20	7.10	9.25	10.35
5.0×10^{-4}	-3.30	4.69	4.70	4.34	3.40	7.10	8.95	10.08
3.3×10^{-4}	-3.48	4.70						
2.0×10^{-4}	-3.70		4.82	4.48	3.55			9.65
1.0×10^{-4}	-4.00		4.90	4.55				
7.5×10^{-5}	-4.12			4.62				
5.0×10^{-5}	-4.30	4.95					7.75	

*Samples heated 5 days at 65°C

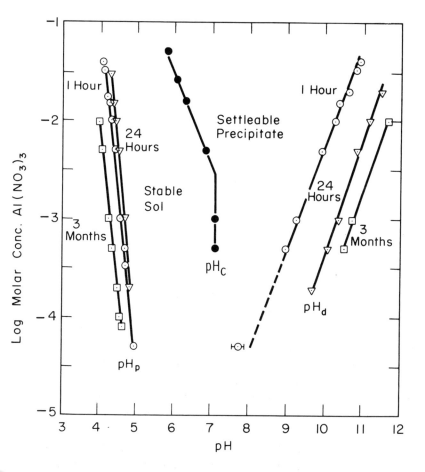

Figure 9.4. Solubility and colloidal stability boundaries of aluminum hydroxide as a function of pH and applied aluminum nitrate concentration.

riangles and 3-month data by squares. The pH-concentration imits of solubility and colloidal stability of the aluminum hydroxide precipitate formed in aluminum nitrate solutions were defined by drawing straight lines through the critical pH values. The slope, intercept and related statistics were calculated using a least squares computer program and are listed in Table 9.4.

Table 9.4
Summary of Statistical Data for Precipitation Boundaries in Aluminum Nitrate Solutions

Critical pH	Slope (S_S)	Intercept (S_I)	R
pH_p (1 hr)	-3.29 (0.01)	12.03 (0.05)	-0.996
pH_p (24 hr)	-3.75 (0.02)	14.41 (0.09)	-0.992
pH_p (3 mo)	-3.57 (0.02)	12.23 (0.07)	-0.996
pH_d (1 hr)	1.05 (0.01)	-12.80 (0.11)	0.991
pH_d (24 hr)	1.14 (0.02)	-14.78 (0.18)	0.990

S_S is one standard deviation in slope, S_I is one standard deviation in intercept, R is correlation coefficient.

In general, precipitation of aluminum hydroxide was detected over wide ranges of aluminum nitrate concentrations and pH. The limits of precipitation were easily defined at the higher concentrations but very precise, sensitive experimental techniques were required at the very low aluminum nitrate concentration. Lower limits of detection were approximately 5.0×10^{-5} M. The pH-concentration region of aluminum hydroxide solubility was defined by the experimental pH_p and pH_d limits of precipitation and dissolution determined at each aluminum concentration. The lines drawn through the pH limits of colloidal stability further divided the pH-concentration region of solubility into two regions (see Figure 9.1). A region of precipitate stability was defined in the acid pH range where a colloidal aluminum hydroxide hydrosol formed and a region of instability was defined in the alkaline pH range where the flocculent precipitate is settleable. Below 2.82 $\times 10^{-3}$ M, the boundary of coagulation was independent of the aluminum(III) concentration. Above this concentration the boundary shifted to more acid pH with increasing aluminum nitrate concentration. This boundary shifted to lower pH and changed slope with time (not shown).

After one hour, the pH boundary of precipitation which defined the limits of aluminum hydroxide solubility in acid solutions shifted to a slightly higher pH. This was apparently

Hydrolysis and Precipitation of Aluminum(III) 341

due to the dissolution of the precipitate formed in the oversaturated solutions during preparation. The shift was greater in the more concentrated aluminum solutions. After 24 hours, the precipitation boundary shifted in the opposite direction, to the acid side, in a rather uniform manner over the entire concentration range. The pH boundary of precipitation in solutions aged three months appears to approach an equilibrium value because further aging up to six months did not significantly change the boundary. Very slight, further shifting of the boundary to the acid side was observed in solutions aged for two years. Also listed in Table 9.3 are pH_p values for solutions heated at 65°C for five days. Apparently, heating of aluminum solutions did not age them in the same fashion as observed at 25°C because the pH data points are not linear with respect to the logarithm of the aluminum concentration.

The entire pH boundary of dissolution shifted to higher pH with aging over the entire aluminum concentration range. The rate of change was somewhat dependent on the time of shaking but did not affect the slope of the pH_d boundary. Although not shown in Figure 9.4, there was a lower limit to the region of the settleable precipitate after aging three months. A stabilized precipitate was detected in a series of solutions between pH 7 and 8 at 7.5×10^{-5} M aluminum nitrate, whereas a flocculent precipitate consisting of large particles was observed in solutions at 1.0×10^{-4} M.

The pH limits of solubility and stability for aluminum chloride solutions aged one hour were determined in a manner exactly as described for the aluminum nitrate system. The results are similar to the one-hour aluminum nitrate precipitation system in that the position of the boundaries of precipitation, coagulation, and dissolution overlap although the slopes and intercepts differ slightly [46].

X-Ray Diffraction Studies

The results of the X-ray diffraction study are summarized in Table 9.5 along with pH data. As shown in the table, five sets of solutions were prepared with OH/Al ratios of 2.90, 2.95,

Table 9.5

Effect of Age on the pH and Crystalline Structure of Aluminum Hydroxide Prepared from 0.0378 M Aluminum Nitrate Solutions

OH/Al Ratio	2.90	2.95	2.95†	3.02	3.25
Initial pH	5.00	5.50	5.50	8.00	9.50
1 Day (A)	4.85* Amor.	5.48* Amor.	5.54* Amor.	7.85* Weak P. boehmite	10.64* Mixture P. boehmite Bayerite
1 Week (B)	4.82* Amor.	5.18* Amor.	5.41* Amor.	7.60* Mixture P. boehmite Bayerite	10.74* Mixture P. boehmite Bayerite Gibbsite
1 Month (C)	4.29* Amor.	4.35* Amor.	5.04* Amor.	7.50* Mixture P. boehmite Bayerite	10.79* Mixture Bayerite Gibbsite
Heated (D)	4.05* Strong Gibbsite	4.10* Gibbsite	4.42* Amor.	7.25* P. boehmite	9.47*

*pH of solution after aging.
†Solution also 0.30 M in sodium nitrate.

2.95 (with 0.30 M sodium nitrate), 3.02 and 3.25. The effect of solution age on the crystalline phase was determined by removing one sample of each set at specific time intervals.

The aluminum hydroxide precipitate recovered from aged solutions at OH/Al ratios of 2.90 and 2.95 was amorphous when aged one month at 25°C, but as indicated by intense peaks at 4.84 Å and 4.35 Å, heating at 65°C for five days produced gibbsite. In the set at OH/Al 2.95 with 0.30 M sodium nitrate, the precipitate was amorphous when aged or heated.

As shown by Figure 9.5, crystalline aluminum hydroxide consisting of bayerite and pseudoboehmite formed after aging one week in solutions with OH/Al ratio of 3.02. After aging one month, the bayerite peaks (4.72 Å and 4.35 Å) and the pseudoboehmite peaks (broad peak at approximately 6.32 Å) were much stronger. At OH/Al ratio 3.25, a mixture of pseudoboehmite and bayerite was identified much sooner in the solid phase. As shown in the figure, intense bayerite peaks were indicated after aging one day at 25°C; however, a weak gibbsite peak appeared after aging for one week and was more defined at one month. The broad pseudoboehmite peak also had disappeared after one month. Heating of the solutions prepared at OH/Al ratio 3.02 and 3.25 for five days produced pseudoboehmite characterized by a broad peak at 6.32 Å.

In general, aluminum solutions prepared at OH/Al ratios less than 3 decreased in pH rather slowly at 25°C but upon heating to 65°C resulted in both a more rapid change and lower pH. In those acid solutions where the precipitate was dispersed as a sol, the pH change was more pronounced than where the precipitate had been coagulated by 0.30 M sodium nitrate and had settled to the bottom of the plastic bottles. Both pH and X-ray data indicate that aging is more rapid for precipitates dispersed as a sol. The pH of the solutions prepared at OH/Al ratio 3.25 increased rapidly at first but changed slowly after 24 hours. Heating of this solution did not increase the pH.

Effect of Sulfate

Precipitation in $Al_2(SO_4)_3$ solutions was studied in a manner identical to that described for aluminum nitrate. Figure 9.6 shows the light scattering-pH curves for 5 x 10^{-4} M aluminum sulfate (*i.e.*, 1 x 10^{-3} M aluminum). The pH range of precipitation was determined by extrapolating the two-minute light scattering data because the precipitate was extremely unstable, settling very rapidly. However, over a very narrow pH range at the pH_p, a slight region or "spike" of restabilization occurred. Comparing this figure with Figure 9.1, at the same concentration of aluminum, the effect of sulfate is obviously quite significant.

344 *Aqueous-Environmental Chemistry of Metals*

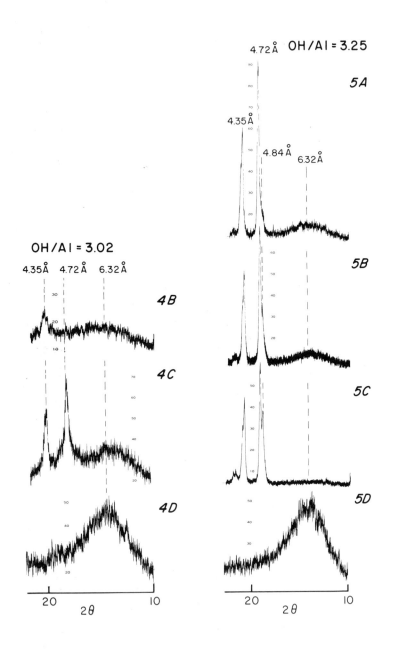

Figure 9.5. Typical results. X-ray diffraction patterns of precipitates recovered from aged aluminum nitrate solutions. **A.** 1 day at 25°C. **B.** 1 week at 25°C. **C.** 1 month at 25°C. **D.** 5 days at 65°C.

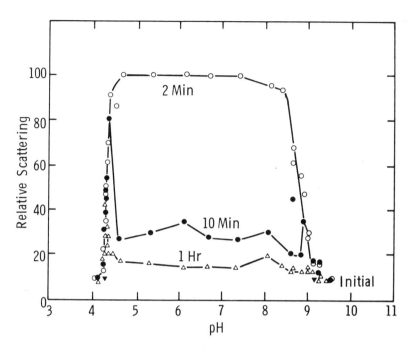

Figure 9.6. Light scattering-pH curves for 5×10^{-4} M aluminum sulfate.

The pH_p decreased 0.30 pH units to lower pH and the aluminum hydroxide hydrosol which was stable in the nitrate system was completely destabilized and settled rapidly in the presence of sulfate. The pH_d was approximately the same as that observed with both the nitrate and chloride solutions.

The pH limits of solubility at different solution ages determined with 5×10^{-4} M aluminum sulfate and other concentrations are summarized in Table 9.6 and plotted as a function of the applied aluminum concentration as shown in Figure 9.7. The symbols are identical to those used for the solubility limits of aluminum hydroxide in Figure 9.4. The slope, intercept, and related statistics for the precipitation boundaries at various aging times are given in Table 9.7. Data on aged aluminum sulfate solutions were obtained as described for aluminum nitrate solutions.

Only the pH_p and pH_d precipitation boundaries are shown in Figure 9.7 because, except for the "spike" near the pH_p, a

Table 9.6
Summary of Critical pH Limits of Precipitation in Aluminum Sulfate Solutions

Molar Conc. $Al(SO_4)_{3/2}$	log M	pH_p (1 hr)	pH_p (24 hr)	pH_p (3 mo)	pH_d (1 hr)	pH_d (24 hr)	pH_d (3 mo)
3.3×10^{-2}	-1.48	3.78					
2.0×10^{-2}	-1.70	3.85			10.71		
1.3×10^{-2}	-1.88	3.88	3.87	3.83	10.55	11.54	11.83
1.0×10^{-2}	-2.00	3.97			10.45		
6.5×10^{-3}	-2.19	4.05	4.00	3.90	10.06	11.26	11.60
4.8×10^{-3}	-2.32	4.11			9.80		
3.3×10^{-3}	-2.48	4.16			9.63		
2.0×10^{-3}	-2.70						
1.3×10^{-3}	-2.89		4.15	4.05		10.33	10.77
1.0×10^{-3}	-3.00	4.25			9.20		
6.5×10^{-4}	-3.19		4.24	4.20		10.14	10.35
4.8×10^{-4}	-3.32	4.40			8.86		
3.3×10^{-4}	-3.48	4.42			8.70		
3.2×10^{-4}	-3.49		4.37	4.30		9.88	10.27
2.0×10^{-4}	-3.70	4.56			8.40		
3.2×10^{-5}	-4.49	4.92				8.25	

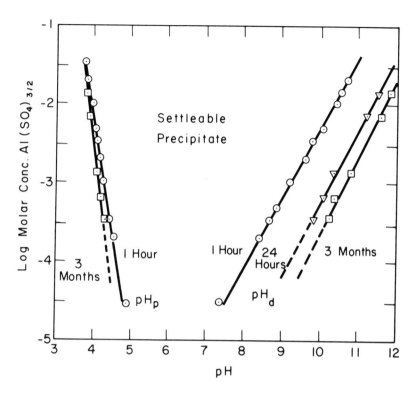

Figure 9.7. Solubility limits of precipitation as a function of pH and applied aluminum sulfate concentration.

Table 9.7
Summary of Statistical Data for Aluminum Sulfate Precipitation Limits

Critical pH	Slope (S_S)	Intercept (S_I)	R
pH_p (1 hr)	-2.97 (0.01)	9.71 (0.05)	-0.996
pH_p (24 hr)	-3.41 (0.04)	11.35 (0.14)	-0.992
pH_p (3 mo)	-3.38 (0.04)	10.99 (0.13)	-0.987
pH_d (1 hr)	0.85 (0.01)	-10.90 (0.07)	0.999
pH_d (24 hr)	0.92 (0.03)	-12.55 (0.28)	0.994
pH_d (3 mo)	0.94 (0.03)	-12.99 (0.38)	0.985

S is one standard deviation and R is the correlation coefficient.

stabilized precipitate did not form at any aluminum sulfate concentration studied. As with the nitrate system, the pH range of precipitation became wider with increasing aluminum concentrations and solution age. The shift of the pH_d line boundary to higher pH was almost identical with that observed with the nitrate system except the slopes were slightly different. The pH_p boundary shifted much less than in the nitrate system, changing only about 0.1 pH units after 3 months with the slope increasing from -2.97 at 1 hr to -3.41 after 24 hrs. Solutions of aluminum sulfate did not appear to be as highly supersaturated—probably because as the precipitate formed, it did not stabilize. For example, the hydroxide to aluminum ratio corresponding to the pH_p was 0.2 at 1 hour and did not change even after 3 months.

The effect of various concentrations of sulfate was also determined. Aluminum nitrate concentration was held constant at 5×10^{-4} M and the concentration of sodium sulfate was systematically varied. The critical pH limits of solubility and colloidal stability determined at each applied sulfate concentration are summarized in Figure 9.8. At the very low sulfate concentrations, at 5×10^{-5} M, the pH limits of solubility and stability are the same as those of 5×10^{-4} M aluminum nitrate. As the concentration of sulfate increased, the pH_p shifted to lower pH; the pH range of precipitation was maximum at 7.5×10^{-3} M sulfate or a stoichiometric aluminum to sulfate ratio of 2:3. The arrow in Figure 9.8 indicates a ratio of 1:1. At sulfate concentrations greater than 10^{-3} M, the pH_p shifted to higher pH in a nonlinear fashion.

At sulfate concentrations greater than 5×10^{-4} M, the precipitate was destabilized across the entire pH range of precipitation except for a very narrow "spike" near the pH_p as described earlier. This region of restabilization formed over such a narrow pH range that it cannot be plotted in the figure. At concentrations greater than 5×10^{-4} M, apparently the positive charged precipitate was coagulated by the sulfate. The nonlinear nature of the pH_c boundary between 0.1 and 0.5 millimolar sulfate suggest both complexation and coagulation.

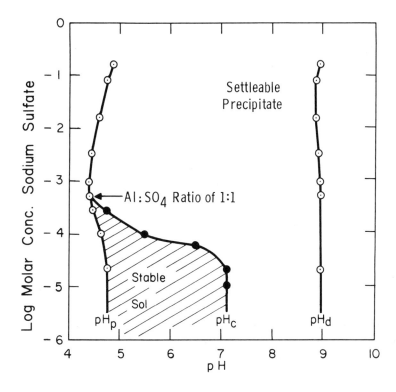

Figure 9.8. Solubility and stability limits of 5 x 10^{-4} M Al(NO$_3$)$_3$ as a function of pH and sodium sulfate concentration.

COMPUTER ANALYSIS OF ALUMINUM HYDROLYSIS

Computer Program SCOGS

The computer has been used extensively by many investigators to solve various chemical equilibrium problems [4,18,30, 47-49]. Most approaches utilize an iterative approximation method to determine equilibrium constants which can be resolved from various forms of input data. The use of a computer presupposes that the results obtained are compatible with the chemistry of the system and that all parameters which may affect the system have been carefully controlled. For example, the concentration of the reactants must be accurately known; and because both temperature and ionic strength affect pH measurements, these parameters must be controlled and held constant. In effect, the results are no better than the experimental data and mathematical approach.

The computer program SCOGS (stability constants of generalized species) is one of several programs formulated by Perrin, Sharma, and Sayce [50-52] which calculates practical or mixed formation constants of complexes with up to two metals and two ligands provided the extent of complexation is pH dependent. The program is based on a mathematical treatment by Wentworth [53,54] and consists of a least squares adjustment to a nonlinear equation. The techniques and results have been reported by Perrin, Sharma, and Sayce primarily for nickel and copper complexes with ligands such as ethylenediamine, histamine and serine.

Originally these programs were designed to determine the equilibrium constants of metal-multiligand complexes. In this study, the original program was modified so that equilibrium constants of large polymeric metal species formed through hydrolysis could be tested. For example, using the unmodified program, the computer cannot handle calculations involving the polymeric ion $Al_8(OH)_{20}^{4+}$. Raising the hydrogen ion activity at pH 4 (1×10^{-4} M) to the 20th power, as is necessary with these species, results in a number beyond the capacity of the computer. To overcome this problem, these calculations were changed to logarithmic form that does not exceed the range of the IBM 360/75 computer.

The input data for the program consist of the molar concentrations of aluminum nitrate, nitric acid, and sodium hydroxide, the pK_w of water at this particular ionic strength and temperature (14.01), the activity coefficient for hydrogen ion calculated by the Davies Equation (0.75), the number and types of proposed complexes, that is, the number of aluminums (m) and hydroxides (q) per complex, and an estimate of the logarithm of the formation constant ($\log \beta_{m,q}$). The volume in ml (V_B) of base added and pH for each data point are also read in. The formation constant $\beta_{m,q}$ must be defined to represent the following hydrolysis reaction and mass action expression

$$mAl^{3+} + qH_2O \rightleftharpoons Al_m(OH)_q^{(3m-q)+} + qH^+ \qquad (13)$$

$$\beta_{m,q} = \frac{[Al_m(OH)_q^{(3m-q)+}]\{H^+\}^q}{[Al^{3+}]^m} \qquad (14)$$

Using the input data, the program sets up two mass balance equations for the metal and ligand concentrations, which in this case are aluminum and hydroxide. The total aluminum concentration which is the sum of the free metal ion and hydrolyzed aluminum concentrations is represented by

$$[Al]_{total} = [Al^{3+}] + \sum_{}^{m}\sum_{}^{q} m[Al_m(OH)_q^{(3m-q)+}] \qquad (15)$$

Solving Equation (14) for $[Al_m(OH)_q^{(3m-q)+}]$ and substituting into Equation (15), the expression obtained is

$$[Al]_{total} = [Al^{3+}] + \sum_{}^{m}\sum_{}^{q} p\beta_{m,q}[Al^{3+}]^m \{H^+\}^{-q} \qquad (16)$$

A similar expression can be written for the total hydroxide activity

$$\{OH^-\}_{total} = \{OH^-\}_{free} + \sum_{}^{m}\sum_{}^{q} q\beta_{m,q}[Al^{3+}]^m \{H^+\}^{-q} \qquad (17)$$

Equations (16) and (17) are the two basic expressions used by the program to refine values for formation constants of the postulated aluminum complexes. Since the total applied aluminum concentration is known, and by fixing the formation constant, the free metal ion concentration can be determined at each data point by Newton-Raphson approximation. Once the concentration $[Al^{3+}]$ is fixed, this value is substituted into Equation (17) and the total hydroxide activity is calculated. From this value, the analytical hydroxide ion concentration is calculated using the activity coefficient (0.75). This quantity is then used to obtain $V_{B(calc.)}$ (ml of base added). The calculated value is then compared with the experimental V_B and their difference, the residual, is determined. As the program progresses through each data point, the least squares equations are built up, and these are solved by matrix inversion to obtain the shifts in the constants. The program proceeds until the sum of the squares of the residuals (ΣR_i^2) is at a minimum. The improved formation constants are then calculated and printed along with their standard deviations. The standard deviation in V_B (S_{V_B}) is also printed out for that particular value of the constant. This cycle is repeated a number of times; usually five are enough to obtain the best fit to the input data.

The modified SCOGS was initially tested using published copper(II) data by Perrin [55] who reported "hand calculated" formation constants for $Cu_2(OH)_2^{2+}$, $p\beta_{2,2}$, to be 10.94. Analysis of the titration data using the modified SCOGS program resolved the constant to be 10.94 ± 0.005, which is in excellent agreement.

Criteria Used for Analysis of Data

In order to determine the nature of the hydrolyzed aluminum complex or combination of complexes present in acid solutions, several statistics calculated by SCOGS were considered. The print-out of this program consists of the refined value of $p\beta_{m,q}$ and its standard deviation along with the standard deviation of the difference between $V_{B(calc.)}$ and V_B. The pH, experimental V_B, residual, total metal concentration (adjusted for dilution) and concentration of each complex were also printed for each data point. Judging from results published by Perrin, Sayce and Sharma [50], SCOGS is capable of refining formation constants with standard deviations of 0.05 or less. The standard deviation of V_B should be approximately the same as the accuracy of the microburet, which for this work was 0.01 ml. The accuracy of the pH measurements in dilute aluminum solutions was limiting. The Sargent-Welch Model DR pH meter is reported by the manufacturer to be accurate within 0.01 pH units, limited principally by the accuracy of the buffers. The incremental volume of sodium hydroxide required to change the pH by 0.01 pH units for the titration data was calculated to be 0.045 ml, which is much larger than the accuracy of the microburet. Therefore a more reasonable standard deviation in V_B would be 0.045 for the aluminum system.

Hydrolysis and Precipitation of Aluminum(III) 353

In this experimental approach, another factor must be considered. When using least squares calculations, each data point is weighed equally and the level of significance increases with the number of data points, but calculations are only meaningful where the concentration of the hydrolyzed aluminum complex is greater than 1 to 2 per cent of the applied aluminum concentration. At concentrations less than 1 to 2 per cent, the pH is not significantly affected regardless of the type of hydrolyzed species, and calculations involving these data points are not particularly significant. Because of this, the data points in the low pH range were not used in refining the equilibrium constants for the complexes with more than one hydroxide per aluminum. Therefore, by assigning an approximate limit to the standard deviation of the logarithm of the formation constant of 0.05 or less and 0.045 to the standard deviation of V_B, the volume of base added, the structure of each hydrolyzed species and its formation constant can be determined with a good degree of certainty.

In addition to the statistics which indicated a certain fit to the experimental data, the results of SCOGS were also plotted so that the various hydrolyzed aluminum complexes under consideration could be compared. Combined hydroxides per aluminum (ligand numbers) were calculated both from the experimental data and the formation constants refined by SCOGS as shown by Figure 9.9. The experimental ligand numbers were calculated at each pH-V_B data point by the following:

$$\bar{n} = \frac{[OH^-]_o - [H^+]_o + [H^+] - [OH^-]}{[Al]_{total}} \quad (18)$$

where $[OH^-]_o$ and $[H^+]_o$ represent the applied concentration of base and acid and $[H^+]$ and $[OH^-]$ represent the free equilibrium concentrations of hydrogen and hydroxide ions. The latter were determined by converting the hydrogen ion activity as measured by the pH electrode to concentrations using an activity coefficient of 0.75. From the formation constants refined with SCOGS, ligand numbers were calculated as a function of pH using

$$\bar{n} = \frac{\sum_{m}\sum_{q} q\beta_{m,q}[Al^{3+}]^m \{H^+\}^{-q}}{[Al]_{total}} \quad (19)$$

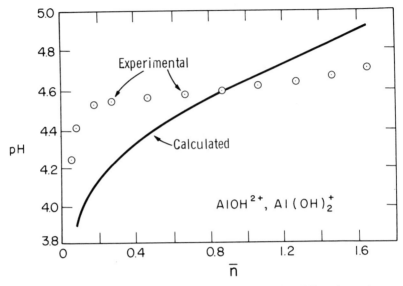

Figure 9.9. Comparison of experimental and calculated ligand numbers as a function of pH. Hydrolysis Scheme 2. 5.0×10^{-4} M $Al(NO_3)_3$, 0.15 M $NaNO_3$. Solutions aged 3 months at 25°C.

Evaluation of Hydrolysis

Six separate and general hydrolysis schemes were postulated and tested in order to determine the soluble hydrolyzed aluminum-(III) complex or complexes in acid solutions below the pH at which the formation of a condensed phase just occurs. These are:

Scheme 1: Al^{3+}, $AlOH^{2+}$
Scheme 2: Al^{3+}, $AlOH^{2+}$, $Al(OH)_2^+$
Scheme 3: Al^{3+}, $AlOH^{2+}$, $Al(OH)_3(aq)$
Scheme 4: Al^{3+}, $AlOH^{2+}$, $Al_m(OH)_q^{(3m-q)+}$
Scheme 5: Al^{3+}, $AlOH^{2+}$, $Al(OH)_2^+$, $Al_m(OH)_q^{(3m-q)+}$
Scheme 6: Al^{3+}, $Al_m(OH)_q^{(3m-q)+}$

The subscripts m and q indicate the number of aluminum and hydroxides per complex. The different combinations of m:q

Hydrolysis and Precipitation of Aluminum(III) 355

examined were 2:2, 2:5, 3:4, 3:7, 4:10, 5:12, 6:12, 6:15, 7:17, 8:20, and 9:23.

The simplest scheme considering only monomeric hydrolysis was tested first. The existence of $AlOH^{2+}$ is generally acknowledged, so the first step was to determine the formation constant $\beta_{1,1}$ for this species at the three concentrations examined. $\beta_{1,1}$ is identical to $*K_1$ used by Sillén and Martel [56]. As this species was to be an important part of five of the six proposed hydrolysis schemes, it was necessary to determine a working value for this constant which could be used to evaluate results obtained in combination with other species. $p\beta_{1,1}$ was refined using only pH-V_B data points from the more acid solutions where hydrolysis would be slight and concentrations of other hydrolyzed complexes would be negligible. Once a value for $p\beta_{1,1}$ was refined, data points at higher pH were added and the existence of $Al(OH)_2^+$ was tested. This combination was tested both by refining values for $\beta_{1,1}$ and $\beta_{1,2}$ simultaneously and then holding $\beta_{1,1}$ constant and refining a value for $\beta_{1,2}$. The results of these calculations are listed in Table 9.8. The values for $p\beta_{1,1}$ of 5.61, 5.38 and 5.51 at 0.1 mM, 0.5 mM and 1.0 mM, respectively, were refined by considering only the more acid solutions where the ml of base added varied from 0.30 to 1.10 for three aluminum concentrations. The low standard deviations in V_B of 0.043, 0.037 and

Table 9.8

Summary of Data for Hydrolysis Schemes 1 and 2

Al^{3+}, $AlOH^{2+}$, $Al(OH)_2^+$

[Al] total	$\log \beta_{1,1}$ $(S_{1,1})$	$\log \beta_{1,2}$ $(S_{1,2})$	Range of V_B	(S_{V_B})
1.0×10^{-3}	-5.61 (0.098)		0.4 - 1.0	(0.043)
5.0×10^{-4}	-5.38 (0.112)		0.6 - 1.1	(0.037)
1.0×10^{-4}	-5.51 (0.098)		0.3 - 1.0	(0.068)
1.0×10^{-3}	-5.50*	-9.04 (0.083)	1.0 - 2.6	(0.330)
5.0×10^{-4}	-5.50*	-9.28 (0.118)	0.9 - 3.0	(0.430)
1.0×10^{-4}	-5.50*	-9.88 (0.098)	1.0 - 2.6	(0.324)

S is one standard deviation

*This value was held constant and not refined.

0.068, respectively, compared to an ideal value of 0.045 indicate a good fit to the experimental data and suggest that $AlOH^{2+}$ is probably the principal hydrolyzed aluminum complex in the more acid solutions. In order to calculate formation constants for other aluminum complexes in solution, an average value of 5.50 was used for $p\beta_{1,1}$. This represents a working value since it may be readjusted once the other species have been determined. The dihydroxo complex $Al(OH)_2^+$ was tested by holding the value of $p\beta_{1,1}$ constant at 5.50 and refining a value for $p\beta_{1,2}$. The results listed in Table 9.8 indicate that $Al(OH)_2^+$ is not an important hydrolyzed aluminum species in acid solutions because the standard deviations in V_B varied from 0.330 to 0.424 which are almost an order of magnitude larger than the ideal of 0.045. Figure 9.9 also indicates a large divergence between calculated and experimental ligand numbers for 5.0×10^{-4} M aluminum nitrate. If the calculated curve closely fits the experimental data, this could indicate that the assumed equilibrium was correct. The smooth curve was calculated using values of $p\beta_{1,1}$ and $p\beta_{1,2}$ of 5.50 and 9.28, respectively.

Hydrolysis Schemes 3 and 4 were tested in a similar fashion. Scheme 3 was tested by refining a constant for $Al(OH)_3(aq)$ while holding $p\beta_{1,1}$ constant at 5.50. The high standard deviation in V_B, for example, 0.242 for 5.0×10^{-4} M aluminum solutions, indicated that $Al(OH)_3(aq)$, like the species $Al(OH)_2^+$, is unimportant in acid solutions if existent at all.

The Scheme 4 series was tested while holding $p\beta_{1,1}$ constant at 5.50 by refining a constant for each of the individual polynuclear aluminum species which have been postulated in the literature or would appear logical. The results for Scheme 4 are listed in Table 9.9. The various values for m, the corresponding number of aluminums, and q, the number of hydroxide ions per complex, are listed in the left columns. For each combination of m and q, the refined value for the logarithm of the formation constant and its standard deviation along with the standard deviation in V_B are listed for 5.0×10^{-4} M aluminum nitrate. The results for five of the polynuclear species tested are also summarized in Figure 9.10. Both the divergence between the experimental ligand numbers (open circles) and calculated ligand numbers (solid line) using constants refined by SCOGS and the standard deviation in V_B are

Hydrolysis and Precipitation of Aluminum(III)

Table 9.9
Summary of Data for Hydrolysis Scheme 4
Al^{3+}, $AlOH^{2+}$, $Al_m(OH)_q^{(3m-q)+}$
$[Al]_{total} = 5.0 \times 10^{-4}$ M

m	q	z	$p\beta_{m,q}$	$S_{m,q}$	S_{V_B}
2	2	+4	4.98	0.791	0.648
2	5	+1	20.07	0.073	0.209
3	4	+5	11.17	0.485	0.543
3	7	+2	25.84	0.100	0.207
4	10	+2	36.39	0.069	0.115
5	12	+3	42.11	0.086	0.118
6	12	+6	38.34	0.212	0.220
6	15	+3	52.55	0.060	0.074
7	17	+4	58.73	0.076	0.081
8	20	+4	68.64	0.053	0.051

V_B = 0.9 - 2.6 ml of NaOH, z = (3m-q), and S is one standard deviation.

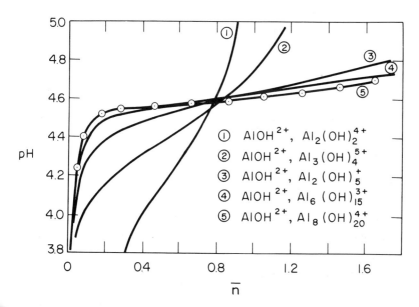

Figure 9.10. Comparison of experimental and calculated ligand numbers as a function of pH. Hydrolysis Scheme 4. 5.0×10^{-4} M $Al(NO_3)_3$, 0.15 M $NaNO_3$.

① $AlOH^{2+}$, $Al_2(OH)_2^{4+}$
② $AlOH^{2+}$, $Al_3(OH)_4^{5+}$
③ $AlOH^{2+}$, $Al_2(OH)_5^{+}$
④ $AlOH^{2+}$, $Al_6(OH)_{15}^{3+}$
⑤ $AlOH^{2+}$, $Al_8(OH)_{20}^{4+}$

criteria which indicate the degree of fit to the experimental data for that particular complex. For example, using a value refined by SCOGS of 4.98 for $p\beta_{2,2}$, the calculated ligand numbers as indicated by Curve 1 diverge sharply from the experimental data. The high standard deviation in V_B of 0.648 also indicates that the dimer $Al_2(OH)_2^{4+}$ is not an important species of aluminum in acid solutions.

The results also discounted most of the polymeric species listed in Table 9.9 as possible aluminum complexes with the exception of three large polynuclear species with an aluminum to hydroxide ratio of 2.5. The results indicated that $Al_6(OH)_{15}^{3+}$, $Al_7(OH)_{17}^{4+}$, and $Al_8(OH)_{20}^{4+}$ could be considered as possible structures because of the low standard deviations in V_B which were 0.074, 0.081 and 0.051, respectively. Although the octamer, as indicated by Curve 5, fit the experimental data very well, the hexamer as indicated by Curve 4 was also considered for further evaluation as well as the septamer.

Hydrolysis Scheme 5 was similar to Scheme 4 except that the $Al(OH)_2^+$ ion was added. The value of $p\beta_{1,1}$ was held constant while $p\beta_{1,2}$ and $p\beta_{m,q}$ were varied one at a time while holding the other constant. All the polynuclear species listed in Table 9.9 were considered and tested. Generally, the results were conclusive in that no combination of $AlOH^{2+}$, $Al(OH)_2^+$ and a polynuclear complex would fit the experimental data. If $\beta_{m,q}$ were held constant and $\beta_{1,2}$ varied, the value of the formation constant for $Al(OH)_2^+$ was so small that if it existed in solution, its concentration would be less than 1 to 2 per cent of the total applied aluminum concentration. If this were true, the presence or absence of this species would not significantly affect the pH of the solution and therefore would be undetectable. For example, the value of $p\beta_{1,2}$ refined in combination with all of the polynuclear complexes with an OH/Al ratio of 2.5 was 25.00 or larger. The standard deviation in V_B was identical to the value refined if $Al(OH)_2^+$ were not included. Some programs were run varying both $\beta_{1,2}$ and $\beta_{m,q}$ simultaneously, but generally the program would diverge rather sharply and would not refine a constant for $\beta_{1,2}$. The constants refined for $\beta_{m,q}$ were consistently almost identical to that refined without considering the single-plus dihydroxy species.

Aluminum hydrolysis excluding mononuclear hydrolyzed species, that is, considering only free Al^{3+} and polynuclear complexes, was tested last. The results of the analysis of Scheme 6 considering the same polynuclear species as before are listed in Table 9.10. Generally the results at the higher pH values are

Table 9.10

Summary of Data for Hydrolysis Scheme 6

Al^{3+}, $Al_m(OH)_q^{(3m-q)+}$

$[Al]_{total} = 5.0 \times 10^{-4}$ M, $V_B = 0.9 - 2.6$ ml

m	q	z	$p\beta_{m,q}$	$S_{m,q}$	S_{V_B}
2	2	+4	5.36	0.487	0.505
2	5	+1	20.13	0.063	0.177
3	4	+5	11.45	0.350	0.418
3	7	+2	25.93	0.080	0.164
4	10	+2	36.48	0.051	0.089
5	12	+3	42.22	0.096	0.096
6	12	+6	38.46	0.189	0.205
6	15	+3	52.69	0.042	0.055
7	17	+4	58.40	0.057	0.063
8	20	+4	68.83	0.040	0.041

z = (3m-q), and S is one standard deviation.

almost identical to Scheme 4, where the various polymeric complexes were tested in the presence of $AlOH^{2+}$. Although the standard deviations in V_B indicated a good fit to the experimental data, when the results were plotted it was obvious that another ionic species must be considered to explain the data at low pH. Although the calculated ligand numbers at the higher pH were almost identical to the experimental values, there was a slight divergence in acid solutions at $\bar{n}$ less than 0.2.

Final Evaluation of Data

Reviewing the several hydrolysis schemes postulated, the only reasonable fits to the experimental data were obtained by

360 *Aqueous-Environmental Chemistry of Metals*

considering the hexamer, septamer or octamer in combination with Al^{3+} and $AlOH^{2+}$. Basically, the final evaluation consisted of first holding $p\beta_{1,1}$ constant at 5.50 and refining a value for $p\beta_{m,q}$. Then $p\beta_{m,q}$ was held constant at the refined value and $p\beta_{1,1}$ was redetermined. The results are listed in Table 9.11. In

Table 9.11

Final Analysis–Aluminum(III) Hydrolysis

Total Al Conc. M	$p\beta_{1,1}$ $(S_{1,1})$	m	q	$p\beta_{m,q}$ $(S_{m,q})$	S_{V_B}
1.0×10^{-4}	5.50*	6	15	52.99 (0.039)	0.050
	5.55 (0.069)	6	15	52.99*	0.049
	5.50*	7	17	58.54 (0.035)	0.041
	5.63 (0.065)	7	17	58.54*	0.040
	5.50*	8	20	68.85 (0.047)	0.048
	5.49 (0.055)	8	20	68.85*	0.048
5.0×10^{-4}	5.50*	6	15	52.55 (0.060)	0.074
	5.84 (0.297)	6	15	52.55*	0.065
	5.50*	7	17	58.23 (0.076)	0.081
	5.71 (0.265)	7	17	58.23*	0.077
	5.50*	8	20	68.64 (0.052)	0.051
	5.61 (0.124)	8	20	68.64*	0.048
1.0×10^{-3}	5.50*	6	15	52.31 (0.037)	0.055
	5.63 (0.163)	6	15	52.31*	0.053
	5.50*	7	17	58.05 (0.050)	0.064
	5.61 (0.185)	7	17	58.05*	0.062
	5.50*	8	20	68.46 (0.029)	0.034
	5.55 (0.105)	8	20	68.46*	0.033
Averages	5.67 (0.176)	6	15	52.45 (0.045)	0.056
	5.65 (0.171)	7	17	58.27 (0.053)	0.060
	5.55 (0.094)	8	20	68.65 (0.048)	0.043

*This value was held constant and not refined.

order to obtain the most statistically significant fit to the experimental data the standard deviations of $p\beta$ and V_B were compared for the three polynuclear species under consideration. Although the standard deviations calculated for the septamer were slightly smaller than those for the octamer at the lowest aluminum concentration considered, 1.0×10^{-4} M, the standard deviations calculated for 5.0×10^{-4} M and 1.0×10^{-3} M were significantly lower for the octamer. The standard deviations in $p\beta_{8,20}$ for the three concentrations were 0.047, 0.052 and 0.029. The standard deviations in V_B were 0.048, 0.048 and 0.033; well within experimental limits of the ideal of 0.045. The standard deviations in $p\beta_{1,1}$ were also smaller when calculated in combination with the octamer. Although polynuclear complexes larger than the octamer could not be tested directly using SCOGS because matrix calculations involved numbers larger than the IBM 360/75 could handle, a larger polymeric species was evaluated by trial and error methods. For example, $Al_9(OH)_{23}^{4+}$ was tested by assuming different values for $p\beta_{9,23}$ and ligand numbers were calculated as a function of pH. The calculated ligand numbers were then graphically compared with the experimental values. The best value used for $p\beta_{9,23}$ was 78.75 but calculations using this value indicated that this species did not fit the data as well as the octamer.

After considering all the data, it was apparent that the only soluble hydrolyzed aluminum(III) complexes which would be present in acid solutions below the pH of precipitation were the monohydroxo-aluminum(III) species, $AlOH^{2+}$, and the octamer, $Al_8(OH)_{20}^{4+}$. The formation constants for these two complexes refined at each aluminum concentration were averaged and the final values were $p\beta_{1,1} = 5.55 \pm 0.094$ and $p\beta_{8,20} = 58.65 \pm 0.043$. The average standard deviations listed above reflect the accuracy of the methods used to calculate the formation constants but do not indicate the experimental aluminum system in that respect. For example, the refined formation constants for the octamer were 68.85, 68.64 and 68.46 for the three aluminum concentrations. Therefore a more realistic value of $p\beta_{8,20}$ would be 68.7 ± 0.2. This value was determined by rounding the constant off to three significant figures and calculating the standard deviation. Obviously the measurements are limiting because more exact formation constants are meaningless. On the other hand, SCOGS is limiting in refining a formation constant for $AlOH^{2+}$ rather than the measurements because the relative concentration of this ion is small compared to the free metal ion and the octamer, and therefore the exact value of $\beta_{1,1}$ is less certain.

Figure 9.11 shows the comparison between experimental ligand numbers (open circles) and calculated ligand numbers (solid line) using the formation constants listed above for $AlOH^{2+}$ and $Al_8(OH)_{20}^{4+}$ at each of the three aluminum concentrations which were studied. The good fit between the calculated and

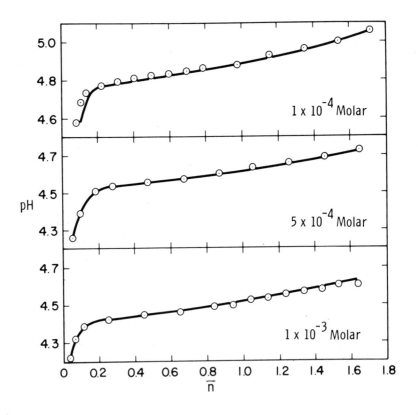

Figure 9.11. Calculated ligand numbers using formation constants for $AlOH^{2+}$ and $Al_8(OH)_{20}^{4+}$ refined by SCOGS. 0.15 M $NaNO_3$, aluminum nitrate solutions.

Hydrolysis and Precipitation of Aluminum(III) 363

experimental data, especially for 5.0×10^{-4} M aluminum nitrate solutions, supports the conclusion that the principal hydrolyzed aluminum(III) complexes in acid solutions are the monohydroxo species and the octameric ion.

DISCUSSION AND CONCLUSIONS

Determination of Equilibrium Constants

In order to study the aqueous chemistry of aluminum(III) in depth, the entire precipitation system must be considered over a broad range of concentrations and pH. Furthermore, this can be accomplished only through a detailed study of both the soluble hydrolysis reactions and the condensed precipitate. The nature and distribution of both the soluble and insoluble hydrolysis products must be determined as each is an integral part and to study one without consideration of the other would lead to unjustified conclusions. It is to this basic premise that this work was directed, a systematic analysis of aluminum(III) precipitation and solubility.

Aluminum nitrate was chosen for the major portion of the work because the nitrate ion does not form soluble or insoluble complexes with aluminum(III) [35,56], an assumption which was experimentally verified [46]. All reactions in nitrate media then can be ascribed exclusively to aluminum(III) hydrolysis.

The exact pH range of precipitaiton at each aluminum concentration was determined at various times after mixing the reactants as shown by Figures 9.1 through 9.3. The limits of aluminum hydroxide precipitation as shown by Figure 9.4 defines the concentration-pH relationship of solubility and colloidal stability. Over the concentration range studied, the pH range of precipitation was well defined using light scattering to detect the presence of solid phase. The concentrated aluminum solutions were visibly turbid requiring less sensitive measurements than the more dilute. Definitive measurements of these very dilute solutions required instrumentation such as the Brice-Phoenix Scattering Photometer. In general, precipitation was detected down to 5.0×10^{-5} M. This is in contrast to reports by Matijević,

Janauer and Stryker [26] and Dezelic, Bilinski and Wolf [57] who found somewhat higher concentration limits of precipitation in the acid pH range. Although there may be other factors involved, this apparent conflict appears to be primarily due to the sensitivity of the instrumentation. If Matijević et al. [26] used the same instrument sensitivity to measure aluminum(III) precipitation as was used to study the coagulation of silver halide sols, then the precipitate in the more dilute solutions would be undetected. For example, the absolute turbidity of a solution of 1.0×10^{-4} M $AgNO_3$ and 5.0×10^{-4} M NaCl (AgCl sol) was determined to be more than one order of magnitude higher than a similar concentration of aluminum(III). Dezelic, et al. [57] measured the turbidity of aluminum solutions after 24 hours. As shown by Figure 9.3, extrapolating 24-hour scattering results at the pH_p could lead to an erroneous pH range of precipitation. A precipitation region similar to these was obtained with the Coleman nephelometer using the relatively insensitive voltage setting of 6.5 volts. Knowledge of the exact extent of the precipitation boundaries is important because several of the conclusions concerning the ionic charge and equilibrium constants of the soluble aluminum species is based on such information.

As would be expected with any hydrolyzing metal, the solubility and stability of the aluminum hydroxide precipitate was strongly pH dependent. The boundaries of the pH_p and pH_d were assumed to be linear so that calculated slopes and intercepts could be used for analysis. It was assumed with some justification [5] that the slopes of the boundaries of precipitation (pH_p and pH_d) are indicative of the charge of the ionic species in equilibrium with the precipitate. The basis of this assumption is that a chemical reaction can be written to describe the ionic equilibrium involved. For example, the aluminate ion $Al(OH)_4^-$ is known to be the ionic species in equilibrium with the precipitate in the alkaline pH range.

$$Al(OH)_3(s) + H_2O \rightleftharpoons Al(OH)_4^- + H^+ \qquad (20)$$

Assuming activity of the solid phase is unity, the mass action expression can be written:

$$^*K_{s_4} = \{H^+\}[Al(OH)_4^-] = \{H^+\} \, C \qquad (21)$$

where C is the applied aluminum concentration. This assumption is valid if $Al(OH)_4^-$ is the only predominant soluble species and if the fraction of the total aluminum present as solid phase is very small in a limiting sense compared to the total applied concentration as it would be along the pH_d boundary of precipitation. Taking logarithms and rearranging into a linear form of y = ax + b, the following relationship is derived

$$\log c = pH - *pK_{s_4} \tag{22}$$

The predicted slope of +1 compares favorably with the experimentally determined slopes of +1.05 at 1 hour and +1.14 at 24 hours calculated using a least squares method.

Assuming a generalized complex $Al_m(OH)_q^{(3m-q)+}$ was the predominant hydrolyzed species in equilibrium with the precipitate along the pH_p, the following equilibrium would apply

$$Al_m(OH)_q^{(3m-q)+} + (3m-q)H_2O = mAl(OH)_3(s) + (3m-q)H^+ \tag{23}$$

The equilibrium constant $*K_{s_{m,q}}$ for this reaction is

$$*K_{s_{m,q}} = \frac{\{H^+\}^{3m-q}}{[Al_m(OH)_q^{(3m-q)+}]} = \frac{\{H^+\}^{3m-q}}{C/m} \tag{24}$$

Taking logarithms and rearranging into a linear form

$$\log C = -(3m-q)pH + (*pK_{s_{m,q}} + \log m) \tag{25}$$

This equation predicts an integer slope of $-(3m-q)$ and intercept of $(*pK_{s_{m,q}} + \log m)$. The calculated slope of the pH_p line boundary at 24 hours was -3.75 which tentatively indicates a 4+ species. Such species which have been postulated are $Al_2(OH)_2^{4+}$ [11,23], $Al_7(OH)_{17}^{4+}$ [18], and $Al_8(OH)_{20}^{4+}$ [27]. Although this method of determining the structure of hydrolyzed species in equilibrium with the precipitate is rather insensitive, it clearly narrows down the possible choices. For example, monomeric hydrolysis leading to precipitation is clearly not indicated. If this were the case the predicted slope would be -1.0 or -2.0.

To determine exactly the structure and formation constant of the polynuclear complex, a modification of the computer program SCOGS was used to analyze potentiometric data. The results indicated conclusively that the experimental data cannot be explained by simple monomeric hydrolysis. The results also discount $Al_2(OH)_2^{4+}$, $Al_3(OH)_4^{5+}$, $Al_6(OH)_{12}^{6+}$, and $Al(OH)_3(aq)$, the latter being soluble aluminum hydroxide molecules. The polynuclear complex showing the best fit to the experimental data over the three aluminum(III) concentrations studied was the octameric ion, $Al_8(OH)_{20}^{4+}$. The identification of the octamer with a 4+ charge using SCOGS correlates very well with the analysis of the precipitation boundaries and provides additional evidence as to the existence of this species.

Several points must be considered in comparing the results of this analysis of aluminum(III) hydrolysis with others. The critical parameters such as temperature, ionic strength, mixing and aging were all controlled; but equally as important, the exact pH of precipitation was determined at each aluminum(III) concentration. Previous investigators have not done the latter. The pH_p for each concentration must be determined since mass balance equations are not valid in the pH range of precipitation because the equivalent concentration of the solid phase cannot be accounted for in the calculations. Potentiometric analysis of solutions in which precipitation has occurred would lead to postulation of unjustified hydrolytic species. For example, Biedermann [18] and Aveston [24] both have reported large polymeric species, $Al_{13}(OH)_{34}^{5+}$ and $Al_{13}(OH)_{32}^{7+}$, respectively. Both analyzed acid aluminum solutions up to OH/Al ratios of approximately 2.5. The results of this work indicate that the presence of precipitate was detected at OH/Al ratios between 0.8 and 2.0 depending on temperature and time of aging. In the presence of precipitate, experimental $\bar{n}$-pH curves flatten out considerably at OH/Al ratios greater than 2.0. Because the shape of these curves is dependent on the number of metal ions per complex and OH/Al ratios, potentiometric analysis of solutions with OH/Al ratios greater than 2.0 would indicate a large polymeric complex which in fact could be a combination of the octamer and aluminum hydroxide. If Biedermann had not included data with the higher OH/Al ratios, the results may have been very similar as he did postulate a septamer $Al_7(OH)_{17}^{4+}$, a species very close to the octamer. The hexamer $Al_6(OH)_{15}^{3+}$ postulated by Brosset [22] as a result of a potentiometric study also correlates fairly well with the octamer.

The results of this work also indicate that $AlOH^{2+}$ must be considered in a hydrolysis scheme; neither Biedermann nor Brosset reported this species. As indicated in Table 9.1, the monohydroxo-aluminum ion is well documented in the literature. Biedermann's results at low OH/Al ratios show a slight deviation between calculated and experimental data which could possibly be explained by including the monomeric species. Since he did not report raw data his results could not be analyzed using SCOGS.

Two methods were used to calculate the magnitude of the formation constant relating the equilibrium between the octamer and aluminum hydroxide. If the slope of the pH_p line boundary were exactly -4.0, the intercept would be exactly (*$pK_{s8,20}$ + log 8). Since the actual slope of the 24-hour pH_p boundary was -3.75, a line was adjusted through the experimental points with a slope of -4.0. The intercept of this adjusted line was calculated to be 15.55. Subtracting log 8 from this, *$pK_{s8,20}$ was determined to be 14.61. The other method consisted of calculating the fractional concentration of the octamer at each experimental pH data point along the boundary using the constants $\beta_{1,1}$ and $\beta_{8,20}$ refined by SCOGS. The stepwise formation constant *$K_{s8,20}$ at each concentration was calculated according to the following

$$*K_{s8,20} = \frac{\{H^+\}^4}{[Al_8(OH)_{20}^{4+}]} = \frac{\{H^+\}^4}{\alpha_{8,20} C/8} \qquad (26)$$

where $\alpha_{8,20}$ is the fraction of total aluminum as the octamer and C is the total applied aluminum concentration. The results of these calculations are summarized in Table 9.12.

The value of p*$K_{s8,20}$ is determined by calculating the concentration of the equilibrium species at each experimental pH_p data point of 14.50 agrees very favorably with the value of 14.61 calculated by determining the intercept of the pH_p boundary assuming a slope of 4.0.

By determining the overall formation constant $\beta_{8,20}$ for the octamer and the stepwise formation constant for aluminum

Table 9.12

Summary of Results for Calculation of $*K_{s_{8,20}}$

(Solutions Aged 24 Hours)

Molar Conc.	$a_{8,20}$	pH_p	$p*K_{s_{8,20}}$
3.0×10^{-2}	0.82	4.22	14.36
1.5×10^{-2}	0.81	4.32	14.46
1.0×10^{-2}	0.82	4.40	14.61
5.0×10^{-3}	0.82	4.51	14.75
1.0×10^{-3}	0.73	4.68	14.68
5.0×10^{-4}	0.57	4.70	14.36
2.0×10^{-4}	0.51	4.82	14.39
1.0×10^{-4}	0.45	4.90	14.35
		Average	14.50 ± 0.17

hydroxide, the overall formation constant for aluminum hydroxide can be calculated using Equations (4), (14) and (26). By taking logarithms, substituting and rearranging, the following relationship can be derived

$$p*K_{s_O} = \frac{p\beta_{8,20} + p*K_{s_{8,20}}}{8} = \frac{68.75 + 14.51}{8} \quad (27)$$

$$= 10.40 \text{ (24 hrs)}$$

$$= 10.05 \text{ (3 mos)}$$

The relationship between the overall formation constant for aluminum hydroxide, $*K_{s_O}$, and the solubility product, K_{s_O}, can also be derived

$$pK_{s_O} = 3pK_w - p*K_{s_O} = 42.03 - 10.40 \quad (28)$$

$$= 31.63 \text{ (24 hrs)}$$

$$= 31.95 \text{ (3 mos)}$$

These values for the solubility product compare very well with constants from the literature listed in Table 9.2.

Assuming there are no other stable hydrolyzed cationic species which would exist only in the presence of precipitate, and therefore undetectable using these methods, the only formation constant not yet determined is for the aluminate ion. Using Equation 22, the calculated p*K_{S4}, assuming exactly a slope of +1.0 for the 1-hour or freshly precipitated aluminum hydroxide, is 12.35 and for the 24-hour system is 13.35. The overall formation constant $\beta_{1,4}$ describing the equilibrium between the aluminate ion and the free metal ion Al^{3+} was calculated by adding p*K_{S_0} and p*K_{S4} and determined to be 23.75 (24 hr).

In order to test the assumptions used in determining the constants, the following formation constants were used to calculate the distribution of aluminum species in a 24-hour system as a function of pH.

$AlOH^{2+}$,	p$\beta_{1,1}$	= 5.55
$Al_8(OH)_{20}^{4+}$,	p$\beta_{8,20}$	= 68.7
$Al(OH)_3$(s),	p*K_{S_0}	= 10.40
$Al(OH)_4^-$,	p$\beta_{1,4}$	= 23.75

Note that these are practical or mixed constants in that hydrogen ion activities as measured with a glass electrode are used along with concentrations of the metal species. $\beta_{1,1}$ and $\beta_{8,20}$ were determined at an ionic strength of 0.15 M. *K_{S_0} and $\beta_{1,4}$ were determined by analysis of the precipitation boundaries, consequently, at varying but minimal ionic strength for that particular aluminum(III) concentration. All were determined at 25°C. The distribution of aluminum(III) species in the absence of the precipitate was calculated according to

$$a_{m,q} = \frac{p\beta_{m,q} [Al^{3+}]^m \{H^+\}^{-q}}{[Al]_{total}} \quad (29)$$

where $a_{m,q}$ is the fraction of total aluminum in that form and not mole fraction.

The distribution of aluminum species in the presence of the precipitate was determined in the following manner. Since the concentration of the free metal ion Al^{3+} is determined by the solubility of aluminum hydroxide, Equation (4) was solved for

$[Al^{3+}]$ and substituted into Equation (29). From this relationship the concentration of the soluble species in equilibrium with the precipitate was calculated. The amount of precipitate removed from solution as the insoluble aluminum hydroxide and the fraction of total aluminum of each aluminum(III) species were calculated as a function of pH. Typical results of these calculations are shown in Figure 9.12. The predominant aluminum fraction over the pH range studied was insoluble aluminum hydroxide. The shapes of the precipitate formation curves are very similar to experimental light scattering curves and the formation and dissolution of the precipitate occurred over a narrow pH range. The calculated pH_p and pH_d values listed in Table 9.13 agree very well with the experimental pH limits of precipitation. In dilute acid solutions below pH 4, the principal ionic species is the free metal ion Al^{3+}. As the pH is increased, hydrolyzed $AlOH^{2+}$ forms to a slight extent, but the soluble

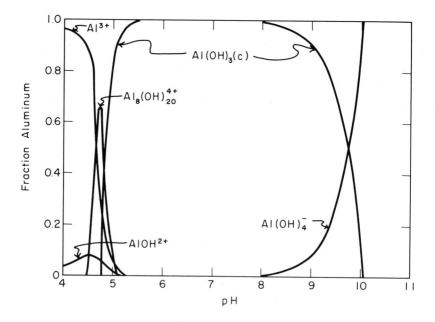

Figure 9.12. Distribution of 5.0×10^{-4} M hydrolyzed aluminum(III) as a function of pH.

Hydrolysis and Precipitation of Aluminum(III)

Table 9.13
Comparison of Experimental and Calculated Solubility Limits of Aluminum Hydroxide Aged 24 Hours

Al(III) Molar Conc.	pH_p (exp.)	pH_p (calc.)	pH_d (exp.)	pH_d (calc.)
3.0×10^{-2}	4.22	4.25		
2.0×10^{-2}			11.47	11.56
1.5×10^{-2}	4.32	4.33		
1.0×10^{-2}	4.40	4.38	11.15	11.24
5.0×10^{-3}	4.51	4.47	10.84	10.88
1.0×10^{-3}	4.68	4.65	10.35	10.33
5.0×10^{-4}	4.70	4.72	10.08	10.05
2.0×10^{-4}	4.82	4.85	9.65	9.59
1.0×10^{-4}	4.90	4.93		

Average Deviation: 0.03 (pH_p) and 0.06 (pH_d)

hydrolyzed octameric species $Al_8(OH)_{20}^{4+}$ is the largest fraction of aluminum present just prior to precipitation. As indicated by the figure, $Al_8(OH)_{20}^{4+}$ is present in appreciable quantities over a fairly narrow pH range. Typical of a polynuclear system, the fraction of octamer is concentration dependent. At 1.0×10^{-3} M Al(III), the peak of the octamer formation curve is 0.78 fraction of total aluminum and 0.53 at 1.0×10^{-4} M. This concentration dependence is also shown in Table 9.12 where the octamer fraction at the experimental pH_p varied from 0.82 to 0.45 over the concentration range studied.

Although the octamer is present in appreciable amounts over a narrow pH range as shown in Figure 9.12, it is present over a wide range of ligand numbers from approximately 0 to almost 3 as is Al^{3+} and $AlOH^{2+}$. The formation curve for the octamer has a peak at OH/Al ratio equal to 2.0 just before the precipitate starts to form and then decreases as the fraction of $Al(OH)_3(s)$ increases. The concentration of all soluble species is minimum at OH/Al of 3.0, which corresponds to the pH_c in a 24-hour system as indicated by Figure 9.4. The calculated range of precipitation of OH/Al ratio from 2.0 to 4.0 agrees with that determined experimentally.

Structure and Form of Aluminum Hydroxide

The light scattering curves constructed as a function of pH indicated that a stabilized precipitate formed in the acid pH range up to OH/Al of 3.0 and a settleable precipitate formed in the alkaline pH range from 3.0 to 4.0. Since the term precipitation has been used rather ambiguously in the literature, the nature of the stabilized precipitate has been in doubt as to whether it is actually a distinct solid phase or in solution as dispersed macromolecules. It is difficult to understand why this conflict has evolved as many other inorganic precipitates are known to form colloidal suspensions and follow the solubility product rule. This difficulty appears to be the result of several factors. First, as indicated by light scattering studies [46], the apparent molecular weight of aluminum hydroxide hydrosol after seven days aging varies over four orders of magnitude ($10^{4.76}$ to $10^{8.58}$ g/mole). If this is true, then the size of the sol particles must also vary in a significant manner over several orders of magnitude. Consequently, depending on instrument sensitivity, the solid phase may be undetected, and it might then be assumed that the products of aluminum(III) hydrolysis are soluble in this pH range. The second factor which has contributed to the theory of soluble large polymers has been the postulation of species such as $Al_{13}(OH)_{34}^{5+}$ and $Al_{13}(OH)_{32}^{7+}$. But as has been pointed out, these species may have been suggested as a result of analysis of solutions oversaturated with respect to the insoluble aluminum hydroxide. The amount of aluminum removed from solution as aluminum hydroxide was not considered in their calculations. Hsu [32] has also suggested a series of large polymeric species which he terms "hydroxy aluminum polymers." He based much of his assumptions on reactions of these polymers in acid solutions and on ion-exchange studies. Hem and Robertson [14] refer to the products of hydrolysis as "aluminum hydroxide polymers" but were uncertain as to the exact nature of the suspended material.

The results of this work indicate rather conclusively that formation of a solid phase occurs over a narrow pH range and the transition from soluble aluminum(III) to insoluble aluminum hydroxide is abrupt and predictable. Due to its size and stability, the precipitate which is colloidally dispersed can be termed an aluminum hydroxide hydrosol. Analysis of the precipitation boundaries yielded equilibrium constants which are logical and consistent with published solubility constants if all the hydrolyzed species are considered. The calculated pH limits of precipitation agree very well with experimental results assuming that the solubility product rule is valid. In addition, molecular weight studies of the aluminum hydroxide hydrosol indicate that the formation and growth of sol particles occur over a narrow range of pH and OH/Al ratios. Therefore it must be considered that the word precipitation can be correctly applied to this system meaning "formation of insoluble aluminum hydroxide."

The data indicate that the predominant factor in the aging process is the evolution of the initially amorphous precipitate to more insoluble crystalline phases. Aluminum solutions in the acid pH range below OH/Al of 3.0 decrease in turbidity initially, due to dissolution of the large particles formed during preparation, but then slowly become more turbid. Heating at 65°C appeared to accelerate the aging process affecting both the turbidity and pH. Long term aging was required to effect the same change at 25°C. The apparent molecular weight of the aluminum hydroxide sol particles increased several orders of magnitude and precipitation was detected at OH/Al of 0.75 and above compared to 2.0 and above at 25°C. The increase in turbidity (*i.e.*, in particle size) during aging is a result of several factors. As the solid phase evolves from the initially amorphous highly soluble state, the concentration of potential determining polymeric aluminum species decreases. This has been shown by Smith [58]. If the concentration of the octamer would decrease on aging, the slope of the precipitation boundary should decrease. This was observed in the aluminum nitrate system where the slope of the pH_p line boundary decreased from -3.75 at 24 hours to -3.57 at 3 months. The decrease of the polymeric species would lead to a reduction in the repulsive surface charge and slow coagulation of the primary sol particles into larger aggregates. The growth of particles could also be as a result of Ostwald Ripening where larger particles grow at the expense of smaller particles. Both mechanisms would promote a decrease in free energy by reducing the total surface area at the solid-liquid interface. Neither of these mechanisms, however, would appear to conform to the classical concept of "olation-oxolation" where the decrease in pH is due to splitting of protons from bound hydroxide ions.

In the alkaline pH range between OH/Al ratio 3.0 to 4.0, the process appears to be more straightforward. The flocculent precipitate which is formed in this region settles rapidly because it is not stabilized by potential determining ions and therefore not colloidally dispersed. The increase in pH with aging is due to formation of a more insoluble form of aluminum hydroxide at the expense of the soluble aluminate ions. The pH change occurs only in the solutions with OH/Al less than 4.0 in which there is precipitate. This would indicate that both aluminum hydroxide and aluminate ions are required for a pH change possible through a surface reaction, $i.e.$,

$$Al(OH)_3(s) + Al(OH)_4^- = 2\ Al(OH)_3(s) + OH^- \quad (30)$$

A mixture of pseudoboehmite and bayerite formed fairly rapidly depending on the initial pH or OH/Al ratio. A weak gibbsite peak appeared in an X-ray diffractogram of precipitate recovered in basic solutions at OH/Al of 3.25 after 7 days and strengthened at 1 month along with a strong bayerite peak. The mixture of gibbsite and bayerite is probably nordstrandite [31]. It would appear that heating would accelerate the aging process as it did in the acid solutions, and gibbsite which is the most stable of the aluminum hydroxide crystalline phases at these temperatures would form. But heating at 65°C did not produce gibbsite but pseudoboehmite, one of the least stable phases.

The pH changes accompanying the formation of more stable insoluble phases is indicated by the shifting of the pH_d boundary to higher pH. The $p*K_{s_4}$ for the aluminum hydroxide-aluminate ion equilibrium varied from 12.35 for a freshly precipitated system to 13.80 for solutions aged 3 months at 25°C with bayerite as the principal crystalline phase. These values bracket the values noted in the literature. This change in the equilibrium constant makes it difficult to assign a specific value to a certain crystalline aluminum hydroxide phase. This may be one of the reasons why there are so many different values listed in the literature [56]. The exact value of pK_4 at a certain time of aging also depends on shaking which was found to accelerate the pH change.

Effect of Anions on Precipitation

Nitrate ion, in excess amounts, did not have an effect on pH limits of aluminum precipitation indicating no significant complexing or ion-pairing to an extent which affects precipitation [46]. Certain anions, however, can affect the precipitation of a hydrolyzing metal in several ways. Although Marion and Thomas [35] pointed out the various effects of anions on the pH of optimum precipitation, their observations can be applied to the pH limits of precipitation. If the anion complexes with the metal and is displaced by hydroxide ions forming aluminum hydroxide, the pH_p will shift to higher pH indicating a higher concentration of hydroxide is required to displace the anion from the soluble complex. If the anion is incorporated into the precipitate forming a mixed salt, the pH_p will be shifted to lower pH.

Sulfate at relatively low concentrations was found to affect both the pH range of precipitation and the colloidal stability of the precipitate. Figure 9.7 shows the precipitation domain of aluminum sulfate and Figure 9.8 shows the effect of varying the sulfate concentration while holding the concentration of aluminum nitrate constant at 5.0×10^{-4} M.

In acid solutions, sulfate forms a soluble complex or ion pair $AlSO_4^+$ [30,56], and has been postulated to form sulfatohydroxo-aluminum(III) species of the general form $Al_x(SO_4)_y(OH)_z^{(3x-2y-z)+}$. Matijević and Stryker [28] have suggested the species $Al_8(SO_4)_5(OH)_{10}^{4+}$. Sulfate is also incorporated into the precipitate at the lower pH range of precipitation forming a mixed salt [32,35]. The results of the precipitation studies support the contention that soluble and insoluble sulfatohydroxo-aluminum(III) species are formed. For example, the slope of the 24-hour pH_p line boundary of precipitation in the aluminum sulfate solutions as shown by Figure 9.7 of -3.41 indicates a 4+ charged species along with species of lower charge. The shift of the pH_p line boundary to lower pH as compared with the aluminum hydroxide boundary as shown in Figure 9.4 indicates that sulfate is incorporated into the precipitate. This would indicate the following equilibrium, assuming the octameric species suggested by Matijević and Stryker.

$$Al_8(SO_4)_5(OH)_{10}^{4+} + 4H_2O = Al_8(SO_4)_5(OH)_{14}(s) + 4H^+ \qquad (31)$$

Unfortunately, this equilibrium cannot be verified experimentally using the techniques developed in this work. In order to study metal-anion complexing by computer analysis of potentiometric data, the anion must hydrolyze to a certain extent. In other words, there must be a pH change accompanying metal-anion interaction. Sulfate is not a hydrolyzing anion as such (K_1 = infinity, K_2 = 1.9) and, in addition, precipitation is initially detected after 0.2 hydroxides per aluminum have been added. This would leave a very narrow pH range in which to study hydrolysis. It is apparent, however, that there is a polynuclear sulfatohydroxo-aluminum(III) species in equilibrium with the precipitate. Assuming $AlOH^{2+}$, $Al_8(OH)_{20}^{4+}$ and $AlSO_4^+$, calculations indicated that the concentration of the octamer was insignificant compared to $AlSO_4^+$ and this species with a +1 charge is not the intermediate species to precipitation or else the slope of the pH_p line boundary would be -1.0.

Figure 9.8 shows the effect of sulfate on the precipitation of aluminum(III) in a most significant manner. At concentrations of sulfate below 2.0 x 10^{-5} M there was no effect on the precipitation limits of aluminum hydroxide. As the concentration of sulfate was increased, the pH_p shifted to lower pH indicating incorporation of sulfate into the precipitate. At approximately $Al:SO_4$ ratio of 3:2 the pH range of precipitation was the widest. Above this concentration the pH_p shifts to higher pH. This indicates that higher pH is required to displace possibly one or more sulfate ions from the soluble complex before a precipitate is formed.

Sulfate also coagulated the stable aluminum hydroxide hydrosol at low concentrations. Above approximately 10^{-4} M the pH_c shifted to lower pH rather abruptly. The nonlinear boundary indicates both aluminum-sulfate complexing and coagulation, the hydrosol being completely destabilized at an $Al:SO_4$ ratio of 1:1.

Inferences to Chemical Coagulation

Although not examined directly, there are several implications of this work to the aggregation of sols with hydrolyzing metals. For example, the 24-hour solubility limits for aluminum-(III) as shown in Figures 9.4 and 9.7 agree with the coagulation limits for various sols reported by Rubin, *et al.* [59-61] and Matijević, *et al.* [26-28]. Because of the pH range over which aluminum(III) is insoluble, the results support the suggestion of Packham [62,63] that coagulation, especially as applied to water treatment processes, is caused by the enmeshment of the sols in the precipitating metal hydroxide. This process is known to chemists as "gathering" [1].

Also, it would appear that hydrolyzed species such as $Al_8(OH)_{20}^{4+}$, which stabilized the hydroxide precipitate, stabilize the various sols as well. It was found that the precipitate is positively charged at pH 7 and below (see Figure 9.2) and it has been established that the sols in this region also assume a positive charge [27,64]. Apparently, the exact position of the restabilization boundary in the sol system depends on the properties and surface concentration of the particular sol [61]. The formation of the boundary, then, as indicated by the c.s.c., would be a function of the equilibrium between the adsorbed species and those in solution as the more indifferent the surface to adsorption, the higher the concentration of free hydrolyzed ions would have to be before restabilization would occur. Coagulation studies in this laboratory with the clay illite indicate that restabilization closely follows the formation of precipitate. Consequently, some colloidal suspensions may require the presence of precipitate before restabilization would occur.

Peterson and Bartow [65], Black and Rice [66], Hanna and Rubin [60], and Packham [63] have shown that sulfate widens the pH range of coagulation. The effect of sulfate on the precipitation of aluminum(III) is shown in Figure 9.8. Although it appears that polynuclear sulfatohydroxo-aluminum species, such as the $Al_8(SO_4)_5(OH)_{10}^{4+}$ ion suggested by Matijević and Stryker [28], may be present, the predominant effect is the coagulation of the forming precipitate by sulfate ions.

Summary of Hydrolysis Constants

The principal ionic aluminum(III) species detected in aqueous solutions were Al^{3+}, $AlOH^{2+}$, $Al_8(OH)_{20}^{4+}$, and $Al(OH)_4^-$. Practical equilibrium constants were calculated by computer analysis of potentiometric data and by analysis of pH limits of precipitation and dissolution over a wide range of aluminum concentrations from 5×10^{-5} M to 5×10^{-2} M at 25°C. Equilibrium constants of insoluble aluminum hydroxide were also determined for the various crystalline phases. The values are:

$*K_1 = \beta_{1,1} = [AlOH^{2+}] \{H^+\}/[Al^{3+}]$

$\log *K_1 = -5.55 \pm 0.09$, Ionic Strength - 0.15 M

$\beta_{8,20} = [Al_8(OH)_{20}^{4+}] \{H^+\}^{20}/[Al^{3+}]^8$

$\log \beta_{8,20} = -68.7 \pm 0.2$, Ionic Strength - 0.15 M

$*K_{S_0} = [Al^{3+}]/\{H^+\}^3$

$\log *K_{S_0} = -10.40$ (aged 24 hrs, amorphous)

$\phantom{\log *K_{S_0}} = -10.05$ (aged 3 mos, gibbsite)

$K_{S_0} = [Al^{3+}] \{OH^-\}^3$

$\phantom{K_{S_0}} = -31.63$ (aged 24 hrs, amorphous)

$\phantom{K_{S_0}} = -31.98$ (aged 3 mos, gibbsite)

$*K_{S_4} = [Al(OH)_4^-] \{H^+\}$

$\log *K_{S_4} = -12.35$ (amorphous, freshly precipitated)

$\phantom{\log *K_{S_4}} = -13.35$ (aged 24 hr, bayerite and pseudoboehmite)

$\phantom{\log *K_{S_4}} = -13.80$ (aged 3 mo, nordstrandite)

REFERENCES

1. Latinen, H. A. *Chemical Analysis.* (New York: McGraw-Hill, 1960).
2. Scoog, D. A. and D. M. West. *Fundamentals of Analytical Chemistry.* (New York: Holt, Rinehart and Winston, 1963).
3. Bilinski, H., M. Branica and L. G. Sillén. *Acta Chem. Scand.,* 20, 853 (1966).

4. Sillén, L. G. and N. Ingri. *Acta Chem. Scand.*, *16*, 159 (1962).
5. Furedi, H. in *The Formation and Properties of Precipitates*, A. G. Walton, Ed. (New York: Interscience, 1967).
6. Fiat, D. and R. E. Connick. *J. Amer. Chem. Soc.*, *90*, 608 (1968).
7. Bronsted, J. V. and K. Z. Volquartz. *Z. Phys. Chem.*, *134*, 97 (1928).
8. Hartford, W. H. *Ind. Eng. Chem.*, *34*, 920 (1942).
9. Schofield, R. K. and A. W. Taylor. *J. Chem. Soc.*, *1954*, 4445 (1954).
10. Ito, T. and N. Yui. *Chem. Abs.*, *48*, 5613 (1954).
11. Kubota, H. *Diss. Abs.*, *16*, 864 (1956).
12. Frink, C. R. and M. Peech. *Inorg. Chem.*, *2*, 473 (1963).
13. Raupach, M. *Aust. J. Soil Res.*, *1*, 36 (1963).
14. Hem, J. D. and C. E. Roberson. "Form and Stability of Aluminum Hydroxide Complexes in Dilute Solution," U.S. Geol. Survey Water-Supply Paper 1827-A, Washington, D.C., 1967.
15. Gayer, K. H., L. C. Thompson and O. T. Zajick. *Can. J. Chem.*, *36*, 1268 (1958).
16. Reesman, A. L. *Amer. J. Sci.*, *267*, 99 (1969).
17. Sillén, L. G. *Quart. Rev.*, *13*, 146 (1959).
18. Biedermann, G. *Svensk Kem. Tidskr.*, *76*, 362 (1964).
19. Sillén, L. G. *Acta Chem. Scand.*, *8*, 299 (1954).
20. Sillén, L. G. *Acta Chem. Scand.*, *8*, 318 (1954).
21. Brosset, C. *Acta Chem. Scand.*, *6*, 910 (1952).
22. Brosset, C., G. Biedermann and L. G. Sillén. *Acta Chem. Scand.*, *8*, 1917 (1954).
23. Faucherre, J. *Bull. Soc. Chem. France*, *21*, 253 (1954).
24. Aveston, J. *J. Chem. Soc.*, *1965*, 4438 (1965).
25. Matijević, E. and B. Težak. *J. Phys. Chem.*, *57*, 951 (1953).
26. Matijević, E., G. E. Janauer and M. Kerker. *J. Colloid. Sci.*, *19*, 333 (1964).
27. Matijević, E., K. G. Mathai, R. H. Ottewill and M. Kerker. *J. Phys. Chem.*, *65*, 826 (1961).
28. Matijević, E. and L. J. Stryker. *J. Colloid. Interface Sci.*, *22*, 68 (1966).
29. Matijević, E., S. Kratohvil and J. Stickels. *J. Phys. Chem.*, *73*, 564 (1969).
30. Stryker, L. J. and E. Matijević. *J. Phys. Chem.*, *73*, 1484 (1969).
31. Schoen, R. and C. E. Roberson. *Amer. Mineral.*, *55*, 43 (1970).
32. Hsu, P. H. and T. F. Bates. *Min. Mag.*, *33*, 749 (1964).
33. Hsu, P. H. *Soil Sci. Soc. Am. Proc.*, *30*, 173 (1966).
34. Hsu, P. H. *Soil Sci.*, *103*, 101 (1967).
35. Marion, S. P. and A. W. Thomas. *J. Colloid. Sci.*, *1*, 221 (1946).
36. Thomas, A. W. and T. H. Whitehead. *J. Phys. Chem.*, *35*, 27 (1931).
37. Thomas, A. W. *Colloid Chemistry*. (New York: McGraw-Hill, 1934).
38. Pokras, L. *J. Chem. Ed.*, *33*, 152 (1956).
39. Pokras, L. *J. Chem. Ed.*, *33*, 223 (1956).
40. Pokras, L. *J. Chem. Ed.*, *33*, 282 (1956).

41. Azabo, Z. G., L. J. Czanyi and M. Kavai. *Z. Anal. Chem.*, *146*, 401 (1955).
42. Raupach, M. *Aust. J. Soil Res.*, *1*, 28 (1963).
43. Frink, C. R. "Reactions of the Aluminum Ion in Aqueous Solutions and Clay Suspensions," PhD Thesis, Cornell Univ., 1960.
44. Schull, K. E. and G. R. Gutham. *J. Amer. Water Works Assoc.*, *59*, 1456 (1967).
45. Larson, T. E. and A. M. Buswell. *Ind. Eng. Chem.*, *32*, 132 (1940).
46. Hayden, P. L. "Aqueous Chemistry of Aluminum(III) and the Solubility and Colloidal Stability of its Precipitates," Ph.D. Thesis, Ohio State University, Columbus, (1971).
47. Hem, J. D. "Graphical Methods for Studies of Aqueous Aluminum Hydroxide, Fluoride and Sulfate Complexes," U.S. Geol. Survey Water-Supply Paper 1827-B, Washington, D.C. (1968).
48. Swinnerton, J. W. and W. W. Miller. *J. Chem. Ed.*, *36*, 485 (1959).
49. Beck, M. T. *Chemistry of Complex Equilibria.* (London: Van Nostrand-Reinhold, 1970).
50. Perrin, D. D., I. G. Sayce and V. S. Sharma. *J. Chem. Soc.*, *1967*, 1755 (1967).
51. Perrin, D. D. and V. S. Sharma. *J. Chem. Soc.*, *1968*, 446 (1968).
52. Sayce, I. G. "Computer Calculations of Equilibrium Constants of Species Present in Mixtures of Metal Ions and Complexing Agents," Australian National University, Canberra, Australia (unpublished).
53. Wentworth, W. E. *J. Chem. Ed.*, *42*, 96 (1965).
54. Wentworth, W. E. *J. Chem. Ed.*, *42*, 162 (1965).
55. Perrin, D. D. and I. G. Sayce. *Talanta*, *14*, 833 (1967).
56. Sillén, L. G. and A. E. Martell. "Stability Constants of Metal Ion Complexes," Special Pub. No. 17, The Chemical Society, Burlington House, London (1964).
57. Dezelic, N., H. Bilinski, and R. H. Wolf. "Precipitation and Hydrolysis of Metallic Ions. IV. Studies on the Solubility of Aluminum Hydroxide in Aqueous Solution," Zagreb University, Yugoslavia (unpublished).
58. Smith, R. W. "Relations Among Equilibrium and Nonequilibrium Aqueous Species of Aluminum Hydroxy Complexes," *Preprints— American Chemical Society, Division of Water, Air and Waste Chemistry*, *10*, 159th National Meeting, Houston (1970).
59. Rubin, A. J. and G. P. Hanna. *Environ. Sci. Technol.*, *2*, 358 (1968).
60. Hanna, G. P. and A. J. Rubin. *J. Amer. Water Works Assoc.*, *62*, 315 (1970).
61. Rubin, A. J. and T. W. Kovac. in *Chemistry of Water Supply, Treatment and Distribution*, A. J. Rubin, Ed. (Ann Arbor, Mich.: Ann Arbor Science Publishers, 1974).
62. Packham, R. F. *J. Appl. Chem.*, *12*, 556 (1962).
63. Packham, R. F. *J. Colloid Sci.*, *20*, 81 (1965).

64. Black, A. P. and S. A. Hannah. *J. Amer. Water Works Assoc.*, *53*, 438 (1961).
65. Peterson, B. H. and E. Bartow. *Ind. Eng. Chem.*, *20*, 51 (1928).
66. Black, A. P. and O. Rice. *Ind. Eng. Chem.*, *25*, 811 (1933).

INDEX

INDEX

absorption spectrophotometry
see atomic absorption
activation control 304
activity coefficients 203,221, 251,350
adsorption 33,36,49,54,60,106,235
adsorption density 45
adsorption free energies 48
adsorption isotherm 45,235
advective transport 131
aeolian transport 131
aeration 194
agriculture 90
air-water interface 35
alkalinity 197,295
alum coagulation 187,318,377
alumina 32,59,132
aluminum 10,11,13,66,179,317-381
aluminum hydroxide 48,319, 342,372
aluminum(III) hydrolysis 11,317,349
aluminum(III) precipitation 317,328,332
aluminum sulfate 342,347
amorphous iron oxide 43
anodic processes 302
anodic stripping voltammetry 171,173,180-186
anthropogenic increase in trace metals 119,137

antimony 5,6,169,174
arsenic 6,169,170,174
atmospheric fallout 86,88
atomic absorption 134,171,173,174,177-179
AutoAnalyzer 175,176
back mix reactors 214
barium 81,169,174,179
bayerite 324,326,342,343,374,378
benthos 152
beryllium 102,168,169,170,174,179
biochemical reactions 106
biogeochemical cycle 3,67
biogeochemical phenomena 3
biological activity 68
biological influence 68
biosphere 67
bismuth 169,174
boron 169,174,179
buffer capacity 200,204,206
bulk diffusion 224
cadmium 3,5,6,7,18,19,20,102, 131,168,169,174,180-188,243,255-289,291-315
  electrolytic deposition 297
  equilibrium chemistry 256
cadmium amine complexes 259,261
cadmium chloride
  complexes 259,260
cadmium corrosion 305
  effect of pH 305

385

cadmium corrosion kinetics	301	contamination	90
cadmium hydroxide complexes	260	continuous stirred-treatment-	
cadmium oxalate trihydrate	243	reactor theory	194,212
calcite	227	copepods	156
calcium	12,21,38,53, 122,179,220-250	copper	5,6,7,12,13,19,30,31,32, 79-122,131,169,174,180, 183,186,187,292,352
calcium carbonate	20,220,225,243		
saturation	294	coprecipitation	33
calcium phosphate	20,220	corrosion	292
calcium sulfate	220,244,331	corrosion inhibitors	306
calgonite	298	corrosive attack by oxygen	293
carbonates	186,197,225, 262,278,295,309,313	critical concentration	222
		crystal growth	219,225
carcinogenic substances	170	crystal growth rate constant	223
cardiovascular diseases	168,169	crystallization	80,222
cathodic depolarization of cadmium	304	crystal surface area	236,251
		Davies equation	203,221,350
cathodic processes	293,302	Debye-Hückel equation	221
charge reversal	48	decomposition of organic matter	194
chelating agents	9		
chelating resins	172	dicalcium phosphate	244,252
chelation with organic ligands	21	dielectric coefficients	51
chemical corrosion inhibitors	306	diffusion	224
chemical interferences	177,178	dislocation theory	241
chemical reactors	193	distribution diagram	259
chemical state	163	dolomite	238
chemical states of mercury	150	domestic wastes	6,93,140
chemical weathering	3	double layer potential	38,40
chromium	7,9,10,53,79-122, 131,169,174,179	dry fallout	90
		EDTA	21,181
clay minerals	34,80	electrical double layer	46,48,50
coagulant	239	electrolyte solutions	7
coagulation	41,318,377	electroplating	270,297
cobalt	6,9,12,19,38,39,41,42, 44,45,53,79-122,168,169,174,180	equilibrium models	2,255
		ethane-1-hydroxyl-1, 1-diphosphonic acid	252
cold vapor atomic absorption	135		
complexation and distribution diagrams	22,31,42,56,185, 197,260,261,264,370	N,N,N',N'-ethylenediamine- tetra(methylenephosphonic acid)	243,252
		ferric oxide	82,194
concentration factors	159	fertilizers	93
of aquatic organisms	107	filtration	194
concentration ratios	156		

flame emission spectroscopy 171
flameless atomic
    absorption 135,173,174
fluid flow models 212
fluvial transport 131
food chain 156
fossil fuels 5,86
free energy 61
fulvates and fulvic acids 106
gallium 81,169
galvanic corrosion 292
galvanic coupling 310
galvanic process 293
gasoline 142
geochemical cycle 130
geothite 43
germanium 169,174,179
gibbsite 43,324,326,
    342,343,374,378
hafnium 12,48
heavy metals pollutants
    impact on ecosystem 162
heterogeneous nucleation 224
homogeneous nucleation 224
Hudson River 119
    metal content 120
humic acids 106
hydrated cation 51
hydrated radius 51
hydrolysis reactions
    11,38,47,187,318
hydrolyzable metal 36,42,49,187
hydroxyapatite 20
hydroxy complexes of cadmium 256
indicator organisms 158
industrial discharges 103
industrial wastes 6,140,256
inhibition of crystallization 228
inorganic forms of mercury 152
inorganic ligands 9,20,194,256
insect larva 156
interfacial processes 3,66

interference mechanisms 177
ion exchange 33,47,60,106
    using chelating resins 172
ionic strength 47,203,251,283
ion selective electrodes 171,174
iron 19,29,37,39,42,53,79-122,
    142,169,174,179,186,193-217
iron hydrolysis 12
iron in oligotrophic lakes 113
iron oxidation 193
iron solubility 196
irrigation water 92
Kaolin 43,48,188
kinetic analysis 171,173,175
kinetic modeling 219
kinetics of crystal growth 219
kinetics of iron oxidation 194
kinetics of precipitation 256,270
labile supersaturated
    solutions 222,224
Lake George, New York 109
Lake Ontario 118
Lake Washington 129,133
land clearing 100
Langelier Index 295
Langmuir equation 53,236
Langmuir-type adsorption 235
lanthanum 48,81
large volume water sampler 132
leachate 7
lead 3,5,6,7,12,18,19,20,24-26,
    37,39,40,48,129-165,
    169,170,174,180,186,187,292
    in biota 153
    doubling time 140
lead alkyls 142
ligand complexes 16,54
light scattering 328,332
lime-soda softening 187
lithium 81
magnesium 21,38,122,179,
    220,221,239,250

magnesium hydroxide 239,243
manganese 19,79-122,169,173,179
manganese dioxide 43,59,82,104
manganite 43
mass transfer 302
mathematic models 12
matrix effects 177,178
mean residence time 144,214
mercury 3-7,17,20,21,22,24-29,
    55-69,129-165,169,170,174
  in biota 153
mesotrophic-eutrophic lakes 114
metal assimilation by organisms 106
metal complexing agents 106
metal distribution in
    natural waters 108
metal ion speciation 181,184,185
metals
  in biota 153
  in coal 83
  in leaves 84,86
  in marine sediments 82
  in oil 83
  in rock 79
  in soils 80,84
  in streams 122
metapolyphosphates 309
metastable region 222
metastable supersaturated
    solutions 222
mine drainage 98
Minnesota lakes 111,117
mobilization 5
molybdenum
    90,169,174,175,176,179
montmorillonite 43,65
multinuclear or polymeric
    complexes — see
    polynuclear hydrolyzed species
natural streams 7,121
Nernst equation 293
neutron activation analysis
    171,172,173

nickel 6,7,12,19,
    79-122,131,169,174
nitrilotri(methylenephosphonic)
  acid 243,252
nordstrandite 326,374,378
nucleation 222,240
NTA 20
ocean sediments 79
olation-oxalation 327,374
oligotrophic-mesotrophic
  lakes 109,113
one-dimensional diffusion 224
organic colloids 35
organic forms of mercury 152
organic ligands 9,194
organic-metal complexing 83
overall formation constant 320
oxidation kinetics of iron(II) 194
oxidation-reduction reactions 10,23
oxide/water interface 36
oxygenation kinetics of
  iron(II) 194,204,206
particulate lead 147
particulate mercury 152
peak current 181
peak potential 181
pH control 197
phenomenological models 46
phosphates 16,93,247
phosphonates 233,239,247
phosphonate additives 243
phosphonate ions 245,299
physical transport processes 68
pluvial transport 131
point of zero charge 43
polarography 171
pollution 90,118,162
polymerization 46
polynuclear hydrolyzed species
    15,37,46,48,320,325,378
polyphosphates 228,229
precipitation
    46,87,88,118,255,318,328

pseudoboehmite	326,342,343,374,378	solubility and solubility product	17,196,261,320,347
Puget Sound	129,133	solubility diagrams	31,42,56,197,222,339,347
pyrite	98	solubility product	318
quartz	37-43,55-59	solubilization	104,105
rate of crystal growth	223	soluble iron	193
reaction rate analysis	173,175	soluble lead	147
reactions of metals in aquatic systems	103	solute-solvent interactions	8
recrystallization	240	sorption effects	181
redox environments	24	sorption of cadmium by kaolinite	188
relative residence time	66	spark emission spectroscopy	168,171
removal processes	130	specific adsorption	48
residence time	66	spectral interferences	177
rivers	118	spectrophotometric methods	171,172
rock and soil	78	spontaneous precipitation	222,225
rubidium	81	stability constants	15,185
runoff	92,93,96	stable supersaturated solutions	225
rutile	43	strontium	81,169,186,187,243
salt water	148	strontium oxalate monohydrate	243
sample collection	132	strontium sulfate	243
Saratoga Lake, N.Y.	114	sulfides	6,10,16,29,31,32
scale formation	295	sulfuric acid	98
scandium	79,81,83,88,90,121	supersaturation	196,201,212,221
scanning electron micrographs	236,238	surface diffusion	224,241
secondary nucleation	223	surface properties	46
sediments	29,148,194	surface reaction	223,224
sediment cores	137	surface sediments	144
sedimentary rocks	80	surface water	92
sedimentation rate	137,138	temperature effects on corrosion rates	294
sediment/water interface	29	thermodynamic ion association constants	220
selenium	168,169,170,174	thermodynamic solubility products	221
sensitivity of analytical methods	171	thorium	12,48
silica	32,41,43,56,59,60	tin	5,6,131,169,174
silicate	309	titanium	79,80,83,88,89,121,169,174,179
silver	5,6,30,32,102,169,174,243		
silver chloride	243	titanium dioxide	80,82
sinks	148		
sodium tripolyphosphate	228,252		
solid/solution interface	33,50		
solid/water interface	36,43		

toxicity of trace elements	169	vanadium	79-122,131,169,174,179
trace metal enrichment	143	wet fallout	90
trace metal history	137	X-ray diffraction	331,341
trace metal removal	186,187	yttrium	81
trace metal sources	3,76	zero point of charge	43,335
trace metal transport	130	zinc	3,5,6,7,18,19,20,30,31,69,
transition metals	77,86,88		79,80-122,169,174,179,180,
transport budget	143		186,187,255,291,292,312,313
transport processes	63	zirconium	81,169,174,179
tungsten	179	zooplankton	156